MANAGING EMOTION IN DESIGN INNOVATION

MANAGING EMOTION IN DESIGN INNOVATION

Amitoj Singh

CRC Press
Taylor & Francis Group
Boca Raton London New York

CRC Press is an imprint of the
Taylor & Francis Group, an **informa** business

CRC Press
Taylor & Francis Group
6000 Broken Sound Parkway NW, Suite 300
Boca Raton, FL 33487-2742

Printed on acid-free paper
Version Date: 20130520

International Standard Book Number-13: 978-1-4665-6750-4 (Hardback)

Library of Congress Cataloging-in-Publication Data

Singh, Amitoj.
Managing emotion in design innovation / Amitoj Singh.
pages cm
Summary: "This book presents an emotion centered research framework titled "emoha" for design innovation. It defines emoha and underlines the importance of the developed framework in culturalization of technology and thereby design innovation. The book explains the detailed research on product styling which leads to the creation of "Emoha" and how to use it in product design"-- Provided by publisher.
Includes bibliographical references and index.
ISBN 978-1-4665-6750-4 (hardback)
1. Product design--India--Case studies. 2. Motorcyclists--India--Psychology. 3. Motorcycles--Design and construction. 4. New products--Psychological aspects. 5. Consumer behavior. 6. Emotions. I. Title.

TS171.4.S56 2013
658.5'7520954--dc23

2013018417

Visit the Taylor & Francis Web site at
http://www.taylorandfrancis.com

and the CRC Press Web site at
http://www.crcpress.com

Dedicated to:

my parents Capt. Gurdev Singh and Harjit Kaur,

my teacher Prof. Lalit Kumar Das,

my elders Ravleen Singh and Kiranpreet Kaur,

my wife Dr. Simardeep Kaur,

and my beloved Indraan Singh, Jisan Singh, Banat Singh & Taarian Kaur.

Contents

List of Figures

List of Tables

Foreword

This book is concerned with emotions. It articulates and details an approach for understanding emotions appropriate to products, user needs, and motivations associated with them in their pursuit. The end objective is to facilitate the users in experiencing the emotions that they relate to their desired product. This can be done by embedding appropriate emotions into the design of products and services. Products with appropriately embedded emotions fit better with the user lifestyle and facilitate satisfaction. A proper design creates a resonance between the user and the product and thereby reduces the dissonance and distortions in the socio-cultural emotional landscape and ecology. This can also create emotional harmony in the society at large.

Designing is a relationship-building activity. Seeking, building, and managing relationships permeate all fields of study. It is also central to synthesis of culture, services, and products. But how are these relationships built? Rationality is one method. Rationality has been the preferred mode of building relationships among scientists, engineers, and technologists. Rationality was at the core of the Modernist movement. It was concerned with building an objective reality, using science and technology. In spite of enormous success, this was soon seen as inadequate and gave way to postmodernism. Postmodernism was based on the position that reality is not objective but is rather constructed, as the mind tries to understand and develop its own personal reality. Postmodernism therefore grew to be sceptical of assertions that one reality fits all. Truth became relative. It became more person and culture specific. In the postmodern understanding, interpretation is a valid proposition; reality only comes into being through our interpretations of what the world means to us individually. The postmodernism era opened the window for emotions to play a role in the development of culture. Gradually, the concept of emotion became respectable.

The use of the term "emotional intelligence" is variously attributed to many authors, Peter Salovey, Jack Mayer, Wayne Payne and others. Later the term was taken as the title for Daniel Goleman's book *Emotional Intelligence: Why It Can Matter More Than IQ*. This book took the message of the significance of emotions vis-à-vis rationality to the world when it became a best seller in 1995. Harvard Business Review published an article on emotional intelligence, which became one of the most popular reprints in its publication history. Emotions, which were looked down upon in management and in the work culture, once again gained respectability.

It was soon discovered that there is a direct correlation between emotional intelligence and effective leadership, team success, and employee performance. Research continues to validate the impact of emotional intelligence in the workplace, as well as creativity. Available assessment tools can indicate individual's emotional intelligence. These tools can facilitate the manpower hiring process, as there is clear evidence that employees more lucid in handling emotions, i.e. having a higher EI, are likely to perform more effectively and contribute to an organization. The concepts of emotional quotients (EQ) and emotional intelligence (EI) thus have permeated into the disciplines of management, design, innovation, and creativity.

The growing interest in emotions is today clearly visible in designing products, services, and sales communications. Although understanding of the importance of emotions in creating an emotional connect with the consumers was well established, no considerable effort was made to evolve method and techniques to design products and services that resonated with the emotional needs and wants of the user. Advertisers have underpinned products and services with emotions and thereby controlled consumer behavior. Products such as automobiles, consumer electronics, lifestyle goods, furniture, and fashion are designed with a great deal of emphasis on emotional considerations. With diminishing technological difference between products, emotions play a significant role in creating unique emotional propositions for products. This influence is seen clearly in emotive styling of products. The service industry too has caught on to the "emotion bandwagon." Companies today realize the importance of creating an emotional connection with users. New and innovative customer service experiences are gaining popularity worldwide.

This book presents a model explaining product-user fit based on the interplay of transactions between the user and the product and the user. The work outlined in the book has wide applicability. It is applicable to all product and services that play a crucial role in facilitating a user's life journey. I have enjoyed being associated with the development of this work. It has immense potential in design of products, services, and corporate identity. I wish the reader a productive journey. We are there to assist. Interaction will help further scholarship.

Lalit Kumar Das
Research Guide
Ex-Head of IDD Centre
Indian Institute of Technology–Delhi
New Delhi

Author's Preface

How to Manage Emotions in Design Innovation

In today's time frame, emotions are paramount to people, and emotional relationships are becoming more and more desirable. An extension of emotional relationships is the affinity we feel with the products that we consume in this physical world. It has become essential to explore a manner in which emotions play a vital role in the physical world with which we surround ourselves. Our ambitions, desires, and needs are all driven by our emotions. The products in the physical world support and propel these aspirations and can be seen as a reflection of our emotional desires. This has implications in the architecture design, interior design, fashion design, design of home appliances, automobiles, and lifestyle products. The physical world can thereby be seen as an extension of emotion-based human aspirations.

Styling is an integral part of any product design and development process. During the styling process, a designer imparts emotive qualities to products. With the diminishing technological differences among products, these emotive qualities play an increasingly significant role in enhancing the desirability and sense of ownership toward the product. However, the absence of quantitative tools to evaluate and determine the emotive quality required in product innovation leads to uncertainty in the styling process.

The Concept of Emoha

This book proposes a new emotion-centered research framework for product styling that can be used for managing emotions in the design innovation process. The developed model has been termed *emoha*-based product styling. The name "emoha" is derived from three words: (1) emotion, (2) *moh* (a Hindi word meaning "attachment"), and (3) "Ha!" (an exclamation of triumph and discovery). *Emoha* refers to the resonance between the emotive needs of a person and the emotive quality of a product and thereby a resonant ownership experience. It is the *emoha-sense* between a person and a product that produces the emotional compatibility that generates a particular product ownership experience. Development of products on the basis of the *emoha-sense* between a person and a product involves this *emoha-based* product styling.

The methodology associated with emoha-based product styling has evolved in the form of a design method. The developed model and design method together represent an emotion-centered research framework for design innovation that can potentially be applied to diverse product categories categorized by cultural segments.

The model divides a product ownership experience into two components: emotive needs for product ownership experience and emotive quality of a product ownership experience. The resonance between these two components creates a mutually reinforcing compatibility between the person and the product. This resonance between the two components has been termed *emoha*.

Motorbikes and Emotions

Motorbikes are a symbol of man's inherent drive to change or become something. It extends our domain of desires and expands our universe. Millions of bikers experience a strong emotional connection with their motorbikes, across the world.

This book is a single case study on motorbikes in India. It develops and validates an emotion-centered research framework for product styling with a case study on motorbikes in India. By providing a systematic way of looking at events, collecting data, analyzing information, and reporting the results, the present study focuses on both generating and testing hypotheses. A mix of both qualitative and quantitative data analysis techniques are used in this book to gain an in-depth understanding of the case under observation. The increasing popularity of motorbikes in India and the writer's prior experience in the motorbike industry were the prime reasons for selecting motorbikes as a case for this study. These factors also helped in maximizing the generation of data in the time period available for the study.

In the present study, though the detailed examination of motorbikes in India cannot provide reliable information about all product categories, it does provide an emotion-centered research framework, which may be tested with other product cases in the future. The emotion-centered research framework developed and tested in case of motorbikes is generalized in the form of a research framework that can potentially be applied to diverse product categories.

This book establishes a segmentation of motorbike ownership experiences in India based on emotive quality. These ownership experiences are seen at the level of *bread-and-butter biking, travel-and-comfort biking, social-networking biking, dexterous and exploratory biking,* and *high-esteem biking* segments.

A biker–bike survey was conducted with 3,106 motorbike owners of 23 different motorbikes to evaluate the biker and bike personality associated with each biking segment. This biker–bike personality signifies the emotive quality of biking. A self-reporting biker–bike personality measurement instrument was developed for the purpose of this study by identifying a cluster of personality variables that characterize motorbikes in India. The results of the quantitative and qualitative analysis established the differentiation in the emotive needs and emotive quality of different biking segments. Styling of products on the basis of this research framework has been termed and explained in the study as *emoha*-based product styling. The approach for using this in the context of a styling studio has been presented and put forth in the book to abet emotion-based design innovation.

Acknowledgements

I am truly indebted to GOD for blessing me through so many kind hearted people who made this book possible.

First and foremost I am grateful to my design guru Prof. Lalit Kumar Das (former Head of IDDC, IIT Delhi) for his invaluable guidance. Without his constant support and encouragement this book could not have been written and published. I will cherish the experience of working under his guidance and carry his values throughout my life. And I am also grateful to him for writing the foreword for this book. This book is dedicated to his teachings.

I am thankful to Dr. Shagun Bhanot (former faculty, IIT Delhi) for her invaluable contribution in proofreading and editing this book. This book was not possible without her professional advice.

I would like to acknowledge with thanks the motivation given by Dr. Keith Russell (The University of Newcastle, Australia) to write this book. I express my gratitude to Prof. B.K. Chakravarthy (Industrial Design Centre, IIT Mumbai) for his guidance.

I am thankful to Dr. Gagandeep Singh, Mr. Jim McGovern, Ms. Laurie Schlags, Ms. Arlene Kopeloff, Ms. Florence Kizza, Mr. Christian Munoz, and the entire team at Taylor and Francis/ CRC Press, for their hard work, patience, and professional support in publishing this book.

I am truly appreciative of the management of IIT Delhi for permitting me to pursue this research. My sincere thanks go to Prof. Amulya Khurana and Prof. Amrit Srinivasan (Department of Humanities and Social Sciences, IIT Delhi) for their guidance and encouragement. My heartfelt gratitude to my colleagues and seniors at Instrument Design Development Centre, IIT Delhi, for their constant support. I am thankful to Dr. D.S. Mehta (IDDC, IIT Delhi) for timely reminding me about the various academic formalities during the course of my research. I also wish to thank Dr. Vigneshwara Illavarasan (Department of Humanities and Social Sciences, IIT Delhi), and Mr. Rajendra LVN (Statistician, Delhi) for their advice and contributions.

Acknowledgment cannot be complete without mentioning Mr. Pawan Kant Munjal, who gave me innumerable learning opportunities during my tenure at Hero Honda Motors Limited (now Hero Motocorp Ltd.). I am extremely grateful to Mr. Brijmohan Lall Munjal for his blessings. My sincere thanks to the management and staff of Hero Honda, especially Mr. Anil Dua, Mr. Sanjay Bhan and Mrs. Anandita Bhandari for their immense support in conducting my research. I am grateful to Mr. Hiroyuki Miyo, Mr. Kamata and Mr. Kamatsu (Honda Motorcycles R&D, India) for being my mentors in the field of motorbike design and research.

I wish to thank all the bikers who spared their valuable time for filling up the questionnaires as well as the entire biking community in India (especially xbp.com), without whom this research would not have been possible.

I am indebted to Mr. Sugato Sen (Society of Indian Automobile Manufacturers, Delhi), Prof. Pieter Desmet (Delft University of Technology, The Netherlands), Prof. Samantha Porter (Loughborough Design School, UK), Dr. Lucas Noldus (Noldus Information Technology bv, the Netherlands), Prof. MariAnne Karlsson (Chalmers University of Technology, Sweden), Ms. Kristine Oustrup (Business Innovation Culture, Singapore), Prof. Andrew Ortony (Northwestern University, USA) and Mr. Sundeep Gajjar (xbhp.com, Delhi) for granting me copyright permissions for using their figures and tables in the book.

I am thankful to Ms. Ayumi Higuchi (Graphic Designer, Amsterdam) for designing the cover for this book and to Mr. Debidutt Gharai (Designer, Delhi) for his illustrations in the book.

I also acknowledge with humble gratitude the contribution of my friends Mr. Mark Wells (Xenophya Design, UK), Mr. S. Sivakumar and Mr. Gopal Krishan Saini (Royal Enfield Motorcycles, Chennai), Mr. Sugandh Malhotra (Studio Ideate, Delhi), Dr. Charru Malhotra (Indian Institute of Public Administration, Delhi), Dr. Chavvi Mehta (Bhavan's Usha and Lakhmi Mittal Institute of Management, Delhi), Mr. Rahul Bhatnagar (Hero Motocorp Ltd., Delhi) and Dr. Gaurav Raheja (IIT Roorkee) who generated various interesting and stimulating discussions during the course of the book.

And last but not the least, I cannot help but acknowledge the unconditional love, support, and the relevant environment provided by my family. I am greatly thankful to my parents, in-laws, elder brother and sister-in-law and my wife and children to whom this book is dedicated.

Dr. Amitoj Singh

The Author

Amitoj Singh is a pioneering PhD holder in the discipline of industrial design from India. His ground breaking research has led to the development of the emotion-centered framework for product design. He is deeply interested in blending his rich industry experience and academic rigor to develop breakthrough and innovative products, services, and systems.

Currently, he is leading product innovation projects at Samsung India Electronics South West Regional HQ. Before joining Samsung, he was working as an assistant professor in the Industrial Design Program at IIT Delhi. During his tenure at IIT he was actively engaged in strategy, research, and innovation consultancy with various global companies in automobiles, consumer products, health care, retail, patent consultants, engineering products, and industrial products, and in the education sectors. Prior to starting his career in academics, he worked as the founding design manager at the industrial design studio (R&D) of Hero Honda Motors Limited. During his tenure of at Hero Honda, he was instrumental in establishing the industrial design studio and leading various motorbike styling projects. His diverse design and innovation projects have brought him acclaim from media and the industry. He has various international publications and patents (applied) to his credit. His specializations include emotion-centered design, design research and strategy, and disruptive innovation management.

1

Introduction

In this introductory chapter, the fundamental concept of the book is augmented through two exploratory studies of **Social Cultural Segmentation of Biking in India** and **Styling Differences within the Biking Segments,** done by the author which lead to an understanding of the core concept of emotion-centered product styling, with a specific emphasis on bikes.

1.1 Importance of Emotions in Product Styling

Styling is an integral part of any product design and development process. During the styling process, a designer imparts emotive qualities to products. With the diminishing technological differences among products, these emotive qualities (or "emotional benefits" as referred by Desmet, 2002) play an increasingly significant role in enhancing the desirability and sense of ownership toward the product. This therefore increases the certainty of the product's success in the marketplace. Knowledge of these emotive qualities would certainly enhance designers' competency in styling products for different ownership experiences.

However, the absence of quantitative tools to evaluate and determine the emotive quality required in products leads to uncertainty in the styling process. Also, the great diversity in people's preferences and product styles makes it extremely difficult to study the emotive quality of each product ownership experience. Therefore, investigating the set of emotive-quality-of-ownership experiences associated with a specific product category in a specific market becomes indispensable to any product styling.

This book introduces a methodology in the form of a framework that can potentially be applied across different product categories in different markets. It is in this context that this work deals with the study of motorbikes in the social cultural environment of the Indian two-wheeler market.

1.2 Need for Studying Emotions Associated with Motorbikes

Vehicles constitute a category of products that give a quantum jump to the social mobility and professional reach of individuals by facilitating the projection of their self to the outside world. In the process, vehicles become a part of users' personality and in turn an expression of their emotions. Motorbikes, being one of the most popular vehicles, reflect the emotions of millions of riders across the world (Figure 1.1).

According to the Automotive Mission Plan 2006–2016 (AMP, 2006), released by the Department of Heavy Industries, Government of India, Asia, with a 90% global market share, is the major producer of motorbikes in the world. Within Asia, China accounts for largest motorbike production followed by India. Out of the 40 million motorbikes that were produced globally in 2005, 7.7 million units were produced in India alone (AMP, 2006). As per the AMP report, the automobile industry is expected to grow from 34 billion USD (5% of the GDP) in 2006 to 145 billion USD (10% of the GDP) in 2016. As shown in Table 1.1, the two-wheelers sales in India grew from 48,12,126 vehicles in 2002–2003 to 74,37,670 vehicles in 2008–2009. More than 88% of the two-wheelers were produced in India in 2008–2009 were sold in domestic markets (Table 1.1).

According to the Society of Indian Automobile Association (SIAM, 2009), two-wheelers (primarily motorbikes) constitute more than 76% of the total

FIGURE 1.1 (See color insert)
Motorbike is an expression of rider's emotions. (From Sunil Gupta of xbhp.com, personal communication, Jan. 27, 2013.)

TABLE 1.1

Automobile Domestic Sales and Production Trend for India

	Automobile Domestic Sales Trends (Number of Vehicles)						
Category	**2002–2003**	**2003–2004**	**2004–2005**	**2005–2006**	**2006–2007**	**2007–2008**	**2008–2009**
Passenger Vehicles	7,07,198	9,02,096	10,61,572	11,43,076	13,79,979	15,49,882	15,51,880
Commercial vehicles	1,90,682	2,60,114	3,18,430	3,51,041	4,67,765	4,90,494	3,84,122
Three-wheelers	2,31,529	2,84,078	3,07,862	3,59,920	4,03,910	3,64,781	3,49,719
Two-wheelers	48,12,126	53,64,249	62,09,765	70,52,391	78,72,334	72,49,278	74,37,670
Grand total	59,41,535	68,10,537	78,97,629	89,06,428	1,01,23,988	96,54,435	97,23,391
	Automobile Production Trends (Number of Vehicles)						
	2002–2003	**2003–2004**	**2004–2005**	**2005–2006**	**2006–2007**	**2007–2008**	**2008–2009**
Passenger vehicles	7,23,330	9,89,560	12,09,876	13,09,300	15,45,223	17,77,583	18,38,697
Commercial vehicles	2,03,697	2,75,040	3,53,703	3,91,083	5,19,982	5,49,006	4,17,126
Three-wheelers	2,76,719	3,56,223	3,74,445	4,34,423	5,56,126	5,00,660	5,01,030
Two-wheelers	50,76,221	56,22,741	65,29,829	76,08,697	84,66,666	80,26,681	84,18,626
Grand total	62,79,967	72,43,564	84,67,853	97,43,503	1,10,87,997	1,08,53,930	1,11,75,479

Source: Retrieved May 12, 2009, from http://www.siamindia.com/scripts/domestic-sales-trend.aspx.

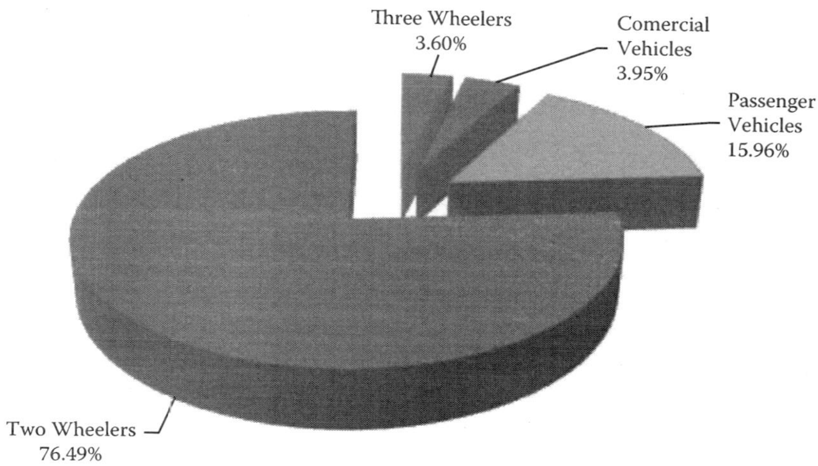

FIGURE 1.2 (See color insert)
Category-wise market share of automobiles in India in 2008–2009. (Retrieved May 12, 2009, from http://www.siamindia.com/scripts/market-share.aspx.)

automobile sales and are therefore seen as significant drivers for economic growth in India at present and in the future (Figure 1.2). Unlike the West, where motorbikes are essentially used for fun and sports, in India these are the most common mode of personal transport. The surge in the number of people with growing aspirations and high purchasing power, along with strong economic growth, over the past few years has attracted major motorbike manufacturers to the Indian market. Companies have put in much-needed investments in R&D, realizing that product design, development, and innovations are crucial for surviving the competitiveness in the industry. With the increase in competition and diminishing technological differences, design rather than technology has become the prime differentiator among various products (Dumanine, 1991; Thackara, 1997, as cited in Desmet, 2002). This is also seen in the case of motorbikes in India. The use of design as a strategic tool for business advantage offers a range of choices to Indian consumers. More than 100 motorbike models are seen on Indian roads, which are primarily differentiated by users on the basis of their aesthetic and emotive qualities in the first instance.

It is seen that while several motorbikes are introduced in the Indian market every year, only a few may evoke sustained emotional attachment among a large number of customers. The motorbike industry in India and in Southeast Asia is essentially a volume-driven industry, and a certain critical sales figure is a prerequisite to justify the expenditure incurred on product development. The failure of a motorbike model to satisfy the needs and aspirations of the prospective owner can critically dent the manufacturer's business objectives. It is for this reason that most unsuccessful motorbike models are often relaunched with frequent styling changes in the hope of enticing the customers and rationalizing the business strategy. Knowledge of "Designing for whom?" rather than "How to design?" has become one of the key parameters to measure a manufacturer's design competency.

1.3 Development of an Emotion-Centered Research Framework

The core focus of this book is to develop an emotion-centered research framework for product styling in the context of motorbikes in India. This is envisioned to assist the design researcher's knowledge and understanding of the emotive quality required in different products for different market segments, by augmenting the manufacturer's design competency.

The following reasons emphasize the significance of developing an emotion-centered research framework for product styling:

1. Most of the time, the design briefs and potential design directions are left to the arbitrary interpretation of designers and market researchers. Similarly, the aesthetic design (sketch, clay model, and prototype model) evaluations too are left to the arbitrary interpretation of the evaluators. The subjective criterion for design selections often leads to conflicts between designers and evaluators over the suitability of a design to meet the business interests.
2. User research studies are at present the focus of the marketing department in most Indian industries and business consultancies. Designers do not possess the appropriate research framework to conduct these on their own. It is for this reason that in spite of participating in user studies, designers often face difficulty in converting the research findings into suitable design concepts. The personal differences in interpreting the research findings also hinder the unblemished dissemination of relevant information to the entire team involved in product development.
3. Though ample equipment and tools are available for designing, simulation, testing, and validation of the objective parameters of design (such as fuel efficiency, drivability, endurance, safety, performance, emissions, etc.), there is a dearth of similar tools to evaluate and determine the emotive content of product styling.
4. Unlike the technical specifications, the emotional concept of a product is not defined on a measurement scale. The lack of a uniform vocabulary or measurement criterion deters the seamless communication of the emotional concept of the product to all stakeholders

who interface with the customers. This often leads to an incongruity between design intent and communication.

5. A trained designer often attempts to accomplish popular styling expressions by taking inspiration from existing products in the marketplace while respecting their intellectual property rights. In spite of being uncertain, this approach may work at times, but it restricts the designer's creativity to a skeleton of predefined styling characteristics. This also makes it difficult for designers to develop styling for unexplored markets.
6. Furthermore, manifold challenges are linked to customer research studies in India. The sociocultural diversity, nonhomogeneous nature of users, and differences in product preferences in different states, regions, cities, villages, and streets as well as in a same household make such studies exceedingly complicated.

The above-mentioned concerns and challenges were personally experienced during the writer's term as a styling designer in Indian motorbike industry. Similar concerns were also shared by other designers working in various automobile and product design industries in India. These concerns provided the prime impetus for developing this emotion-oriented research methodology.

1.4 Research Design and Methodology

The choice of which method to employ depends on the nature of the research problem (Noor, 2008) and upon the type of the social phenomena to be explored (Morgan and Smircich, 1980). In this regard, it is important to study ownership experiences in real-life contexts, as their emotive quality may vary with different product categories and market segments. However, it is extremely difficult to study each ownership experience owing to the large diversity of products as well as the demographic and geographic segmentation of consumers. Also, an experiment-based study of different product categories in a controlled environment would not represent the real-life context. Therefore, rather than conducting experiments with different product categories, the research methodology aimed at a holistic, in-depth investigation of a specific product category in a specific market. It is for this reason that an emotion-centered framework for product styling applicable to designing motorbikes in India was chosen as a research strategy. The developed framework is further generalized in the book for testing its potential applications in other product categories (or cases) as well.

1.4.1 Case Study as a Research Strategy

According to Zainal (2007), case studies explore and investigate contemporary real-life phenomena through detailed contextual analysis of a limited number of events or conditions and their relationships. Thus, a case study research enables the researcher to explore and understand the data within the context of its use (Yin, 1984; Zainal, 2007) and to gain a holistic view of a certain phenomenon (Feagin, Orum, and Sjoberg, 1991; Gummesson, 1991). According to Flyvbjerg (2006), a case study produces context-dependent knowledge that helps transform the researchers from rule-based beginners to virtuoso experts. This book consists of a single case study on motorbikes in India. By providing a systematic way of looking at events, collecting data, analyzing information, and reporting the results, it focuses on both generating and testing hypotheses. A mix of both qualitative and quantitative data analysis techniques are used here to gain an in-depth understanding of the case under observation. The selection of the case of motorbikes is based on information-oriented sampling (Flyvbjerg, 2006). The increasing popularity of motorbikes in India (explained in the previous section) and the writer's prior experience in the motorbike industry were the prime reasons for selecting motorbikes as a case for this study.

According to Abercrombie, Hill, and Turner (1984, p. 34), "though a detailed examination of a single case cannot provide reliable information about the broader class, but it may be useful in the preliminary stages of an investigation since it provides hypotheses, which may be tested systematically with a larger number of cases." Similarly, in the present study, though the detailed examination of motorbikes in India cannot provide reliable information about all product categories, it does provide a research framework that may be tested with other product cases in the future. Generalization of results from case studies stems on theory rather than on populations (Yin, 1984). Therefore, the emotion-centered research framework developed and tested in the case of motorbikes is also generalized in the form of a research framework that can potentially be applied to varied product categories.

1.4.2 Generation of Research Hypothesis

The book combines several exploratory studies with the case study of motorbikes to understand the social reality of biking in India. These exploratory studies were based on a series of field trips, personal interviews, focus group discussions, participatory workshops, photo documentations, discussions with motorbike designers/experts, online discussions on biking forums, motorbike design projects, and biking trips across various cities in India undertaken by the writer during his professional experience in the motorbike industry. The first study led to the development of a theoretical model of six biking segments: *bread-and-butter biking, travel-and-comfort biking, social-networking biking, dexterous and exploratory biking, high-esteem biking,* and *religious biking* to represent the social-cultural segmentation of biking in India. This was followed by a visual comparative study of 10 different motorbike models to investigate the styling differences within the identified biking segments.

The findings of the two exploratory studies established the premise that the identified biking segments can be differentiated on the basis of the associated emotive needs for biking and emotive qualities of biking. In the process, it also came to notice that the conjoint biker–bike personality represents the emotive quality of the bike ownership experience.

To investigate the established premise in greater depth, a self-reporting biker–bike personality measurement instrument was developed. This was done by identifying a cluster of 14 personality-describing adjectives (personality variables) related to motorbikes in India. The development of the instrument was based on online word-count frequency investigation conducted using Inspyder Insite 2.0 word-count software. The investigation was based on a study of 31, 61,092 words extracted from seven motorbike websites in India.

A purposive sampling method was used for selecting the sampling frame. A total of 9,512 motorbike owners of 23 different motorbikes belonging to 24 different states of India constituted the sample population. An online investigation was considered the most suitable data collection method to extend the study.

The data after collection was classified into two broad categories: qualitative and quantitative. Quantitative data included the personality ratings given by the respondents on the 14 personality variables, and the qualitative information was collected from the respondent's feedback to the open-ended question. The qualitative data were analyzed first to ascertain the social-cultural segmentation of biking. This was followed by the quantitative analysis of the biker–bike personality ratings given by the respondents to assess the emotive quality of biking segments.

1.4.3 Generalization of Results

The results of the qualitative and quantitative analyses establish the difference in the configuration of emotive needs and emotive quality associated with each of these biking segments. The premise of the study establishes the segmentation of motorbike ownership experiences in India in the form of five biking segments: *bread-and-butter biking, travel-and-comfort biking, social-networking biking, dexterous and exploratory biking*, and *high-esteem biking*.

The subsequent findings were thereafter correlated and represented in the form of an emotion-centered model for motorbike styling in India. The developed model for motorbike styling was adapted to form a generic model for product styling that has been termed as a model for *emoha*-based product styling. The methodology associated with *emoha*-based product styling was developed in the form of a design method for *emoha*-based product styling. The developed model and design method together represent an emotion-centered research framework for product styling that can potentially be applied to diverse product categories that are categorized by cultural segments.

1.5 Social-Cultural Segmentation of Biking in India

The social reality of our lives is a hard, day-to-day, factual reality of birth, survival, and growth. The products we use often reflect this reality or influence the formation of a social reality specific to the product. The social reality connected with the product represents the context in which the product exists.

The first-hand experience of the writer in motorbike industry during 2002–2008 facilitated exposure to a series of field trips, personal interviews, focus-group discussions, participatory workshops, photo documentations, discussions with motorbike designers/experts, online discussions on biking forums, motorbike design projects, and biking trips across various cities in India. These led to an understanding of the experiences, preferences, aspirations, lifestyles, needs, and motivations of bikers across India. At the same time, the concept of needs, motivations, wants, and desires was reviewed through a preliminary study of the literature. The understanding of "needs" gained from the study of diverse disciplines of economics, psychology, consumer research and marketing, sociology, cultural anthropology, and social philosophy was followed by a deeper study of need theories and typology developed by Galtung (1980), Mallmann (1980), Maslow (1954, 1968), Max-Neef (1991, 1992), and Mc Dougall (1928).

Psychologist Abraham Maslow, known for his concept of hierarchy of needs, suggests that people are motivated to fulfill basic needs before moving on to other needs (Maslow, 1954, 1968). Maslow's hierarchy of needs is most often displayed as a pyramid, with the lowest levels of the pyramid made up of the most basic needs and with the more complex needs at the top of the pyramid. Needs at the bottom of the pyramid are basic physiological needs. Once these lower-level needs have been met, people can move on to the next level of needs, which are safety and security needs, love and belonging needs, esteem needs, and need for self-actualization. Thus, the needs pyramid represents the man's journey of life toward fulfillment of his or her needs. It is important to note that the physical environment (including products) also plays an important role in this journey of a person. The role that a product plays in this journey influences the product ownership experience. The diversity in human needs as well as in a product's ability to satisfy them leads to segmentation of ownership experiences.

In reality, a motorbike cannot be seen in isolation from its biker. A bike and biker together complete a biking or motorbike ownership experience. Biking embodies the relationship between the biker and the bike and is influenced by bikers' emotive needs for purchasing, riding, or owning a motorbike. The heterogeneous needs of bikers have given rise to different social groups of bikers. Motorbikes are a means of satisfying diverse needs for these social groups. Also, each social group is characterized by a common set of attitudes, values, goals, practices, and behaviors toward biking and represents a specific biking culture. The heterogeneous needs of biker groups and the differences in the biking cultures have given rise to a social-cultural segmentation of biking. A clear similarity could be identified between the social-cultural segmentation of biking and the concept of Maslow's hierarchy of needs (discussed above). A theoretical model of six biking segments: *bread-and-butter biking, travel-and-comfort biking, social-networking biking, dexterous and exploratory biking, high-esteem biking,* and *religious biking* was developed to represent the social-cultural segmentation of biking in India (Figure 1.3).

Bikers are not a homogeneous group. The needs and wants of biking may vary among bikers from different countries, cultures, families, and groups. This differentiation related to biking is reinterpreted on Maslow's pyramid (Figure 1.3). In the present study, the concepts of "needs" and "wants" are not differentiated and are hereby referred to as "emotive needs of biking," or "biking needs," in the context of Maslow's hierarchy of needs.

The distinct characteristics of each biking segment are explained in this section. The findings are based on contextual observations.

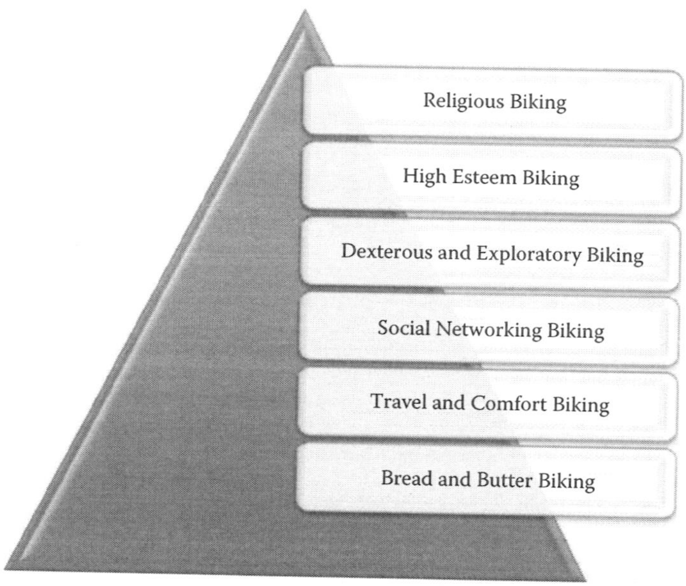

FIGURE 1.3
Social cultural segmentation of biking in India.

1.5.1 Bread-and-Butter Biking Segment

> Hero Honda CD 100 is the first motorbike in my life; only timely servicing is required. This is most comfortable bike for local running as well as long drives.
>
> **(quoted verbatim from male, 54 years, Jharkhand, employed in the public sector)**

The bottom of the pyramid represents all types of two wheelers, which are purchased and used by customers as a means to fulfill their basic physiological needs. Bread-and-butter biking is often connected with professional and commercial applications, and therefore the bikers in this category may often not be the actual owners of their motorbikes. Biking for them is a means of commuting and earning a livelihood. Krige (1995, as cited in Tunnicliff, 2006) in one of his studies on motorcyclists' attitudes and behavior stated that "commuters" ride for practical reasons rather than for the love of riding. The motorbikes belonging to this biking segment are therefore primarily evaluated on rational parameters of cost and ability to perform the desired function.

Figure 1.4 shows a pictorial representation of the bread-and-butter biking segment in India.

FIGURE 1.4 (See color insert)
Pictorial representation of the bread-and-butter biking segment in India. (Illustration by Debidutt Gharai, personal communication, Jan. 28, 2013.)

1.5.2 Travel-and-Comfort Biking Segment

> Today, the motorbike is one of the family members, because (1) it helps us to reach our destination fast and on time, so we can complete the work on time. (2) I strongly agree that it is like a family member because it will help us in emergency situations, for example, suppose at late midnight anything sudden happens to us. (3) I like my bike very much because I feel it is like my wife to me.
>
> **(quoted verbatim from male, 26 years, Andhra Pradesh, businessman)**

The next segment on the pyramid is that of travel-and-comfort biking. This segment represents a slightly more emotionally involved biker who uses the motorbike as a mode of transport for the entire family. Though the bikes are primarily evaluated on rational parameters of safety, comfort, fuel efficiency, convenience, and ease of maintenance, cultural mannerisms also start influencing the user's preference at this level of biking. In the process, the biker starts to develop associations with biking communities having similar emotive needs. Bikers linked with this biking genre start becoming sensitive to the connotation of brand and associated communications as well.

Figure 1.5 shows a pictorial representation of the travel-and-comfort biking segment in India.

1.5.3 Social-Networking Biking Segment

> My passion is for bikes. I read every bike's review and test-drive all bikes released in market. I love to go on tours with friends. It is the public's easy, fast-moving and flexible transport system, which allows me to be omnipresent.
>
> **(quoted verbatim from male, 22 years, Andhra Pradesh, student)**

Brands and advertising play an important role in creating bonds among bikers. Motorbike manufacturers across the globe create brand personalities to develop meaningful brand associations among their followers. These associations help create biking communities. A Harley-Davidson rider can be distinctively differentiated from a Ducati sports enthusiast based on their distinct emotional connection with the respective brands. Social networking biking is all about loving a particular brand and belonging to a group of like-minded brand patrons. This biking segment represents a collective community expression. Cotte and Ritchie (2005) further categorized such a consumer as a "tribal consumer" who is driven by a fundamental need to be accepted as a part of a larger group or a "sophisticated consumer" who is conscious of the advertiser's persuasive intentions and communication tactics.

Figure 1.6 shows a pictorial representation of the social-networking biking segment in India.

FIGURE 1.5 (See color insert)
Pictorial representation of the travel-and-comfort biking segment in India. (Illustration by Debidutt Gharai, personal communication, Jan. 28, 2013.)

FIGURE 1.6 (See color insert)
Pictorial representation of the social-networking biking segment in India. (Illustration by Debidutt Gharai, personal communication, Jan. 28, 2013.)

1.5.4 Dexterous and Exploratory Biking Segment

> I have confidence in my power driving. To see 'mum said no girls' on my bike. To have a look of my power ride on my bike and style.
>
> **(quoted verbatim from male, 30 years, Andhra Pradesh, employed in the private sector)**

This category represents bike enthusiasts who are driven by the emotive need of exploring their self-potential. Biking for them is a means of realizing their creative imagination. Krige (1995, as cited in Tunnicliff, 2006) described such bikers as young and inexperienced "Boy Wonders" who love challenges, race their peers, and often do not consider the consequences of their behavior. The dexterous and exploratory biking lifestyle is often associated with sports and advanced riding skills. Such bikers are constantly looking for improvements in performance, in technology and in the design of their bikes.

FIGURE 1.7 (See color insert)
Pictorial representation of the dexterous and exploratory biking segment in India. (Illustration by Debidutt Gharai, personal communication, Jan. 30, 2013.)

Figure 1.7 shows a pictorial representation of the dexterous and exploratory biking segment in India.

1.5.5 High-Esteem Biking Segment

> I feel like God on the road. Weekend getaways and hangouts. Personal work.
>
> **(quoted verbatim from male, 21 years, Tamil Nadu, employed in the private sector)**

Bikes in this category are often associated with high self-esteem. Manufacturers design and develop flagship premium models to cater to the high-esteem biking enthusiasts (such as the Harley Owners Group). Krige (1995, as cited in Tunnicliff, 2006) described these subgroups as "Weekend

FIGURE 1.8 (See color insert)
Pictorial representation of the high-esteem biking segment in India. (Illustration by Debidutt Gharai, personal communication, Jan. 30, 2013.)

Warriors" because they ride as a hobby and for fun. Bikes are often seen as an expression of self-image and self-esteem by such rider groups.

Figure 1.8 shows a pictorial representation of the high-esteem biking segment in India.

1.5.6 Religious Biking Segment

> My bike moves my soul along with my body. I find my bike to be my best friend. I love bikes.
>
> **(quoted verbatim from male, 33 years, Maharashtra, employed in the private sector)**

At the highest level of Maslow's hierarchy, biking is akin to a religion. At the top of the pyramid, the rider and machine cannot be differentiated. The bike becomes an expression of the self. This level represents the highest degree of personalization in motorbikes.

While personalizing a product's appearance, a person invests energy in the product to express his/her self (Mugge, Schifferstein, and Schoormans, 2004). Research shows that personalization of products leads to greater attachment of a person toward the products (Mugge et al., 2004). According to Govers and Mugge (2004), congruity between the personality of a person and the personality of the product leads to a stronger product attachment. A *religious biker* is passionate about the bike and therefore loves to modify his machine to match his personality. It is due to this reason that in this segment the emotional attachment of a biker with his motorbike is much more than in any of the preceding segments of biking.

Figure 1.9 shows a pictorial representation of the religious biking segment in India.

FIGURE 1.9 (See color insert)
Pictorial representation of the religious biking segment in India. (Illustration by Debidutt Gharai, personal communication, Jan. 30, 2013.)

1.6 Styling Differences within the Biking Segments

According to Pham (1999), style refers to designs that possess a few recognizable common characteristics. The preferences for a particular design can vary among individuals, thereby, making it difficult to typify those designs. However, it is possible to amalgamate varied preferences into the style expressions that characterize the proposed biking segments (Figure 1.3).

The proposed segmentation of biking represents a unique framework for categorizing motorbikes on the basis of the different emotive needs of the bikers. Table 1.2 illustrates the classification of 48 motorbike models of eight different manufacturers in India, on the five biking segments. Since religious biking is associated with personalization and custom biking, it is not included (hereafter) in the present study of mass-manufactured bikes.

A visual comparative study was conducted on 10 different motorbike models of Hero Honda Motors belonging to different biking segments. This study was confined to the evaluation of prominent aesthetic properties of form, graphics, and colors. The objective of the comparative evaluation was to identify the distinct characteristics of form, color, and graphics among bikes belonging to different biking segments. To preclude the influence of brand biases, the writer restricted the study to the users of a single motorbike company. Also, the choice of the company was considered appropriate since the selected manufacturer has more than 55% share in the domestic motorcycle market (Hero Honda, 2009) and a product range that covers all the proposed biking segments.

The 10 motorbikes that were investigated are Hero Honda CD Dawn, Hero Honda CD Deluxe, Hero Honda Splendor Plus, Hero Honda Splendor NXG, Hero Honda Super Splendor, Hero Honda Passion Plus, Hero Honda Glamour, Hero Honda Glamour FI, Hero Honda Hunk, Hero Honda CBZ

TABLE 1.2 (See color insert)

Classification of Indian Motorbikes on the Five Biking Segments

Hero Honda	Bajaj	Honda	TVS	Yamaha	Suzuki	Kinetic	Royal Enfield
CD-Dawn	XCD 125	Unicorn	Star	Alba	Heat	Stryker	Bullet Machismo
CD-Deluxe	Platina	Shine	Star City	G5	Zeus 125	Comet	Bullet Machismo 500
Splendor Plus	Discover	CBF Stunner	Star Sport	Crux	Hayabusa	Aquila	Thunderbird
Splendor Nxg	Pulsar DTS-i		Flame	Gladiator			Bullet Electra 5S
Passion Plus	Pulsar 200 DTS-i		Apache RTR 160	MT-01			Bullet Electra
Super Splendor	Pulsar 220 DTS-Fi			YZF-R1			Bullet 350
Glamour	Avenger DTS-i			FZ1			
Glamour Fi	CT 100			YZF R15			
Achiever							
Hunk							
CBZ Xtreme							
Karizma							

High Esteem Biking
Dextrous and Exploratory Biking
Social Networking Biking
Travel and Comfort Biking
Bread and Butter Biking

Xtreme, and Hero Honda Karizma. Figure 1.10 shows the distinction in forms, graphics, and colors among these motorbikes.

1.6.1 Styling in Bread-and-Butter Biking

The bikes associated with bread-and-butter biking have a distinct geometric form. This can be seen in the fuel tank designs of CD Dawn and CD Deluxe. As explained earlier, the bikes in this segment are purchased and used by customers as a means to fulfill their basic logistic needs and are therefore evaluated on rational parameters of cost and ability to perform the desired function. The mechanical fabricated shape of the fuel tank and other body parts enhance the functional image of these bikes. The graphics on the bike distinctly follows the form as well. Similar to the form, the graphics too are geometric in shape and primarily follow a horizontal axis. However, the graphics are scattered on different body parts and lack any continuity. The graphic pattern is abstract and occupies a significant area on the motorbike. The body color is in monochrome, and the graphics are complementary to the body color.

1.6.2 Styling in Travel-and-Comfort Biking

The Splendor series of motorbikes (Splendor, Super Splendor, and Splendor NXG) belong to the travel-and-comfort biking segment. The forms of these bikes are slightly evolved from the previous geometric forms. The fuel tank is more smooth and curvaceous as compared to the solid geometric tank of CD Dawn. As explained before, this segment of biking represents a slightly more emotionally involved biker who uses the motorbike as a mode of transport for the entire family, and therefore, these bikes are primarily evaluated on parameters of safety, comfort, fuel efficiency, convenience, and ease of maintenance. This is reflected in the curvaceous design of fuel tank, which gives a smooth and comfortable look to the bike as compared to the mechanical image of bread-and-butter biking designs. The graphics of this category adheres to the tank profile and occupies a large area, as in the case of bread-and-butter biking designs. Whereas Splendor has scattered graphics, Super Splendor and Nxg have continuous graphics that run throughout the body. These designs are associated with the maximum number of body colors, and the graphic color schemes are both analogous and complementary to the body color. Solid colors are replaced by a composition of metallic shades and merging gradations.

1.6.3 Styling in Social-Networking Biking

Bikes such as Glamour and Passion Plus characterize social-networking biking, in which a bike is a means to socialize. Brands and advertising play an important role in creating bonds among bikers belonging to this form

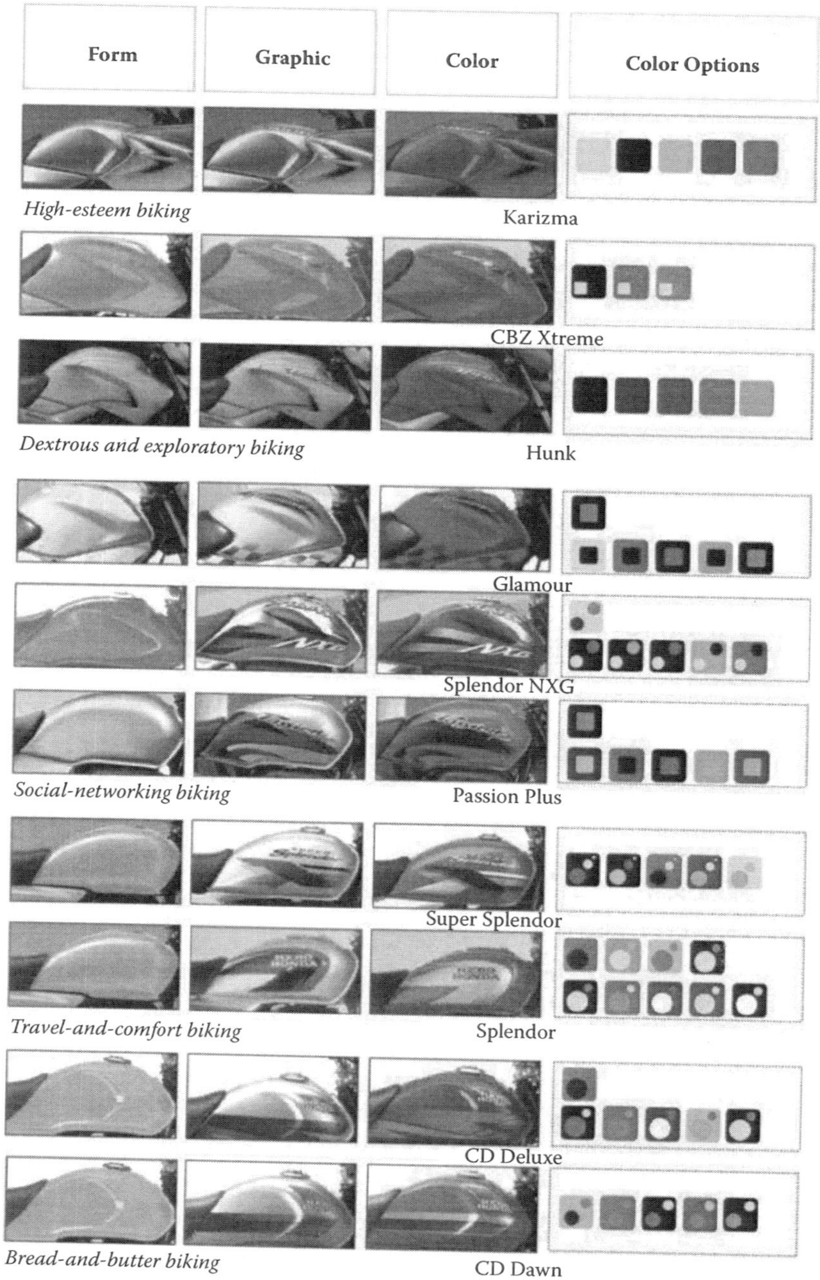

FIGURE 1.10 (See color insert)
Visual comparison of aesthetic characteristics (form, graphics, and colors) of motorbikes representing different biking segments.

of biking. Motorbike manufacturers across the globe create distinct product personalities to develop meaningful brand associations among their followers. The aesthetic properties are thus more distinct and metaphorical in character. Even the names of the bikes belonging to this segment are more emotive (Glamour and Passion Plus) as compared to the bikes belonging to bread-and-butter biking (CD Dawn and CD Deluxe) or travel-and-comfort biking (Splendor, Super Splendor, and Splendor NXG).

The forms are more playful and organic. The presence of additional features such as scoops and knee grips on tank add to the bike's character and distinguishes it from other bikes in its group. Though the tank of Passion Plus is also smooth similar to that of Splendor, the proportions are quite different. Whereas the Splendor tank is more balanced and gives a stable feel, the tank of Passion Plus is more dynamic owing to its larger length-to-height ratio. Unlike the preceding designs, the graphic designs associated with social-networking biking are representational. The graphic patterns on Passion resemble a human tattoo, whereas the Glamour graphics of flames and checkers represent the popular racing culture. An uninterrupted flow of graphics along the body contours gives a distinct character to these designs. Dual tone hues and analogous shades replace the multiplicity of colors seen in the preceding levels of the biking hierarchy.

1.6.4 Styling in Dexterous and Exploratory Biking

The dexterous and exploratory biking lifestyle is often associated with sports and advanced riding skills. Such bikers are constantly looking for upgrades in performance, technology, and design of their bikes. The fuel tanks of the bikes such as CBZ Xtreme and Hunk are shown in Figure 1.10. The forms of these two bikes are quite different. The CBZ Xtreme is represented by sharp lines and edges, whereas the Hunk has a more organic and muscular fuel tank. These bikes have more complex and advance forms as compared to bikes belonging to the preceding biking segments. The distinct forms give a clear and cutting-edge image to the CBZ Xtreme and a powerful and aggressive image to the Hunk. The forms are both creative and exploratory in nature. Graphics are abstract, minimalist, segregated, and in opposition to the body contour. Colors are multitone and single tone with complementary shades.

1.6.5 Styling in High-Esteem Biking

Bikes such as the Karizma, which belong to this biking segment, are often associated with high self-esteem. The forms of bikes in this segment seem to be more confident and distinct and less playful in nature. The presence of distinct curves enhances the identity and esteem of the bike. The graphics are either minimalist or replaced by three-dimensional metal finish emblems. The bikes in this segment are associated with minimum color variations.

1.7 Generation of the Premise of Study

Dichter (1960) stated that knowledge of "soul of things" is possibly a new revolutionary way of discovering the soul of man. At the same time, discovering the soul of man can give insights into his desires and preference for things. In this socially constructed world, man is seen as a product of society (Berger and Luckman, 1966). Therefore, examining the role of motorbikes as an expression of human personality can give insights into the beliefs, values, attitudes, and emotions of the societies and cultures where these motorbikes are created, purchased, or used.

Material culture studies (Appadurai, 1986; Attfield, 2000; Dant, 1999; Douglas and Isherwood, 1978; Prown, 1982; Sahlins, 1976) are based on the premise that objects reflect the beliefs of the larger society to which they belong. The motorbikes belonging to different biking segments should therefore reflect the belief, attitudes, and emotions of the bikers. The identified biking segments were re-examined in the endeavor to investigate the relationship of the material culture of biking with that of the styling of motorbikes.

A deeper probe into the social-cultural segmentation of biking reveals that what differentiates bread-and-butter biking, travel-and-comfort biking, social-networking biking, dexterous and exploratory biking, and high-esteem biking segments is not only the set of biking needs but also the style of the motorbikes (Figure 1.11). A comparison of motorbikes highlighted the difference among biking segments in their choice of aesthetic characteristics. By satisfying the emotive needs of bikers, these aesthetic characteristics should subsequently start reflecting the emotions and beliefs of the biking segment to which they belong. The two exploratory studies discussed earlier suggested a possible differentiation in the emotive qualities of different biking segments.

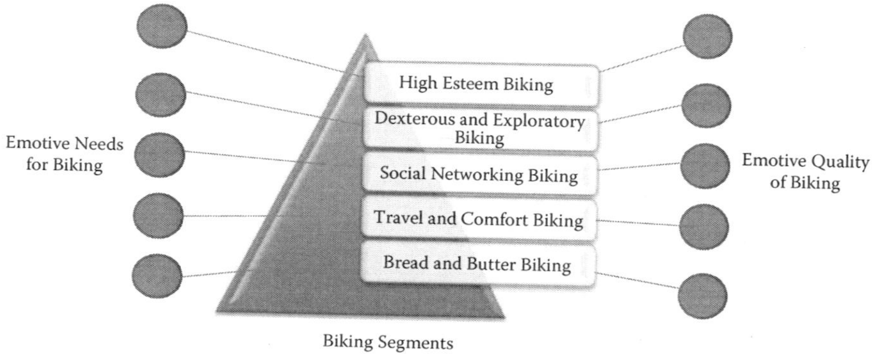

FIGURE 1.11
Visual representation of the premise.

The exploratory studies thereby lead to the generation of the following premise: The bread-and-butter biking, travel-and-comfort biking, social networking biking, dexterous and exploratory biking, and high-esteem biking segments can be differentiated on the basis of the associated emotive needs for biking and the emotive quality of biking.

During the exploratory studies (Section 1.5), it was observed that in social reality a motorbike cannot be seen in isolation from its biker. A bike and biker together complete a biking or motorbike ownership experience. These observations suggested that the emotive needs of biking can be studied among motorbike owners with reference to their current motorbikes owned. And similarly the emotive quality of biking can be studied by evaluating the bike's image or personality in relation to the self-image of the biker. An emotion-centered research framework was therefore needed to investigate and ascertain the differences in the emotive qualities within the different segments of biking in India.

2

Literature-Oriented Research Framework

The review of the literature for developing the focal research framework was conducted in three parts. The first part focused on examining the research traditions in design and emotions, followed by a review of the current approaches for designing emotion-laden products.

The second part of the literature review investigated the theoretical applicability of the research premise generated in Chapter 1.

The last part of the literature review examined the significance of the present study for motorbike design. The role of emotions specific to design innovation in motorbikes was studied in detail. The emotion-based design strategy of major global motorbike manufacturers such as Honda, Yamaha, Harley-Davidson, Ducati, and BMW was examined in relation to their choice of emotions and expressions in design.

2.1 Research Traditions in Design and Emotions

2.1.1 Understanding of Emotions

"After all, we all 'know' what an emotion is. That is, until we are asked to give a definition" (Desmet, 2002, p. 3).

We all experience emotions in our everyday lives. Emotions may be connected to our interactions with other fellow human beings, objects, situations, and even with our inner self. There are times when we know that we are experiencing an emotion, while at other times we may not even be aware of our emotional reaction to a situation. In our lives, we act as a subject or object of emotions while experiencing them ourselves or instigating emotions among others, respectively.

Though the concept of emotions has been widely researched, it is quite difficult to find a commonly acceptable definition. The word *emotion* derives from the Latin verb *emovere,* which means, "to move out" (De Rivera, 1977, p. 11). The *Oxford Advanced Learner's English Dictionary* (OALD, 2005) defined emotion as "a strong feeling such as love, fear or anger; the part of a person's character which consists of feeling." This is one of the simplest definitions of emotion and relates emotions to feelings. Feelings are defined as "something which you feel through the mind or through the senses: a feeling of hunger/excitement/sadness, etc." (OALD, 2005).

2.1.1.1 Theories of Emotions

Hartvig Dahl, in his introductory essay in the book titled *A Structural Theory of Emotions* by De Rivera (1977, p. 1), stated, "If we begin by asking what an emotion *is* we soon find that we are instead asking a whole set of other questions, such as: What does an emotion feel like? Where do we feel it? How do we express it? What apparatus produces it? On and on." It is therefore important to study the theories of emotions in order to comprehend emotion.

De Rivera (1977) divided the diverse theories of emotion into three broad categories: emotions related to instincts, emotion and the perception of value, and transformational theories. The various theories on emotion explain the concept of emotion from different perspectives.

McDougall (1926) related emotion to human biological instincts. According to his theory, an object or a situation can arouse an instinct that results in a behavioral response and an experience of an emotion. For example, a viewer may jump in joy while watching his favorite comedy show on television. The situation here arouses the viewer's instincts, which direct his behavior (to jump) and experience an emotion (of joy). Tomkins (1962), on the other hand, suggested that the affect system indirectly motivates behavior. According to him, the affect system directly invokes bodily expression, which leads to an emotion. This viewpoint suggests that each bodily expression (e.g., facial expression) is specific to an emotion. According to James (1890), emotions are our perception of bodily changes. A perception of a situation leads to an action, and an afferent feedback from our actions results in an emotional experience. Paulhan's (1884) explanation differed from the previous instinctual theories in stating that only a part of instinctual energies that is not used in action manifests itself as an emotion or an emotional expression. Also, this theory does not define a particular order of occurrence for emotion and related bodily expressions.

Arnold (1960) described emotion on the basis of the appraisal of an emotion-inducing situation or an object. According to her, an appraisal of a situation leads to action tendencies that cause emotions and bodily expressions. Unlike McDougall's theory, in which instincts directly affect action, Arnold's action tendencies may only provide a motive for action. Angyal (1941) in his theory explained that the quality of emotions is influenced by both perception of value and bodily feedback. Many researchers (Lazarus, Averill, and Opton, 1970; Mandler, 1975; Schachter and Singer, 1962) supported the appraisal theory of emotions, suggesting that emotions consist of appraisal and arousal.

According to Sartre (1948), in an emotional state a person reacts to an object in a magical nondeterminate world. The magical world causes a transformation in the person–product relationship. In such a situation, certain key qualities of a product become more important than the rest of the object or rest of the logical determinate world. This explains our emotions toward a taillight of a car and our desire to own it without considering its other aspects or other viable competitive options. Thus, emotion occurs when a person is transported from a logical to a magical world. Similar to Sartre, Pribram and Melges (1969) also stated that emotions occur when the flow of action is prevented.

2.1.1.2 Cognitive Structure of Emotions

Emotions are always about something (Gordon, 1974). Averill (1980) also explained this intentional characteristic of emotions. He explains, "A person cannot simply be angry, fearful, or proud. He [or she] must be angry *at* something, fearful *of* something, proud *of* something" (p. 310).

Parkinson and Manstead (1992) explained emotions as a reaction to the personal meanings of situations. Since emotions, as stated earlier, are personal and intentional in character, they can be viewed in many different ways. A situation can trigger different emotions in different people. Memory also plays a significant role in the cognitive process of generating emotions. Hence, multiple factors play a role in eliciting emotional responses. According to Parkinson et al. (1992), emotions are multilevel syndromes that develop over time. The appraisal theory links the concepts of these differentiating emotions to an individual's evaluation of an emotion-inducing situation at any given point in time.

Many theorists have agreed that cognitive appraisals are central to emotion. In order to understand how emotions are generated by an object or a situation, it is important to understand the structure of emotions. The cognitive structure of emotions proposed by Ortony, Clore, and Collins (1988) and the structural theory developed by De Rivera (1977) explained the structure of emotions on the basis of subject–object interaction (appraisal) and the subject–object relationship (transformation), respectively.

Ortony et al. (1988) defined emotions as valence reactions to events, to agents, or to objects and are characterized by the way in which the eliciting situation is construed. According to them, this distinction between reactions to events, agents, and objects gives rise to three basic classes of emotions: "pleased versus displeased" (reaction to events), "approving versus disapproving" (reactions to agents), and "liking versus disliking" (reaction to objects). According to this structural theory, a person's appraisal of an emotion-inducing situation is based on three central variables: desirability, praiseworthiness, and appealingness, which apply to event-based emotions, agent-based emotions, and object-based emotions, respectively.

De Rivera (1977), on the other hand, proposed a structural theory of emotions based on a transformation of relationship between the experience and the object of experience. According to him, the experience of emotion reflects the transformation of our relationship to the world and that each type of emotion (anger, fear, love) reflects a different kind of transformation. De Rivera described different emotions by conceiving particular emotions as movements in a three-dimensional (belonging, recognition, being) interpersonal space. He developed a matrix of object relations in which a particular emotion is delineated by specifying its position in this matrix.

2.1.1.3 Classification of Emotions

Larsen and Diener (1992) explained that some emotions are similar to one another, but yet significantly different from other emotions. There has been an extensive research in the field of emotions by different researchers; researchers have proposed ways of differentiating emotions. This concept of emotional differentiation can be illustrated in Table 2.1, which shows how different researchers have classified fundamental emotions.

TABLE 2.1

A Selective List of Fundamental or Basic Emotions

Theorist	Fundamental Emotions	Reference
Arnold, M. B.	Anger, aversion, courage, dejection, desire, despair, fear, hate, hope, love, sadness	Arnold (1996)
Ekman, P.	Anger, disgust, fear, joy, sadness, surprise	Ekman, Friesen and Ellsworth (1982)
Frijda, N.	Desire, joy, pride, surprise, distress, anger, aversion, contempt, fear, shame	Frijda (1987)
Gray, J.	Rage/terror, anxiety, joy	Gray (1982)
Izard, C. E.	Anger, contempt, disgust, distress, fear, guilt, interest, joy, shame, surprise	Izard (1972)
James, W.	Fear, grief, love, rage	James (1884)
McDougall, W.	Anger, disgust, elation, fear, subjection, tender-emotion, wonder	McDougall (1926)
Mowrer, O. H.	Pain, pleasure	Mowrer (1960)
Oatley, K., and Johnson Laird, P. N.	Anger, disgust, fear, happiness, sadness	Oatley and Johnson-Laird (1987)
Panksepp, J.	Expectancy, fear, rage, panic	Panksepp (1982)
Plutchik, R.	Acceptance, anger, anticipation, disgust, joy, fear, sadness, surprise	Plutchik (1980)
Tomkins, S. S.	Anger, interest, contempt, disgust, distress, fear, joy, shame, surprise	Tomkins (1984)
Watson, J. B.	Fear, love, rage	Watson (1930)
Weiner, B.	Happiness, sadness	Weiner and Graham (1984)

Source: Ortony, A., Clore, G. L., and Collins, A. (1988). *The Cognitive Structure of Emotions*. New York: Cambridge University Press. p. 27.

2.1.1.4 Emotions Associated with Products

Desmet (2002), in his pioneering book *Designing Emotions*, followed the appraisal theories of emotion to propose a model of product emotions (Figure 2.1). The model identified "concern," "stimuli," and "appraisal" as the three variables that lead to an emotion. According to this model, it is the appraisal of a product to satisfy or dissatisfy a person's concern that leads to an emotion.

Desmet further explained that the concerns of a person need to be studied in regard to the context in which one interacts with the product. This explanation helps to understand, in the context of the present study, why bikers belonging to different biking segments may experience different emotions with the same motorbike style. This difference may be primarily due to the distinct concerns of bikers, as explained in the previous chapter. Second, different bikers may also associate distinct meanings with the same motorbike. For example, a chrome-plated cruiser motorbike can elicit different emotions among different bikers. A biker may perceive "a chrome-plated cruiser" as masculine and powerful and thereby experience pleasant emotions (like joy), whereas another biker may perceive it to be jarring and environment-unfriendly and thereby experience unpleasant emotions (like disgust). This, however, assumes that the emotive concern of the first biker is to own a masculine and powerful bike, while a jarring bike does not satisfy the emotive concerns of the second biker. The differentiation in the emotional experiences of the two bikers can be attributed to the differentiation in the appraisal of the motorbike with respect to individual concerns. Desmet not only proposed a model for product emotions but also classified a set of 41 emotions related to products (Figure 2.2).

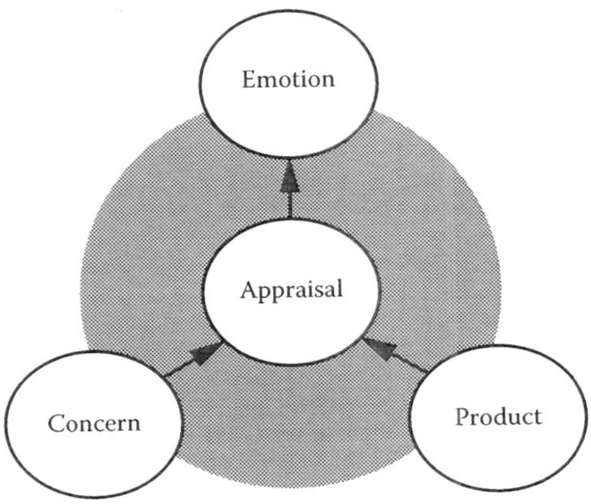

FIGURE 2.1
Desmet's model of product emotions. (From Desmet, P. M. A. (2002). *Designing Emotions*. Delft, the Netherlands: Delft University of Technology, p. 107.)

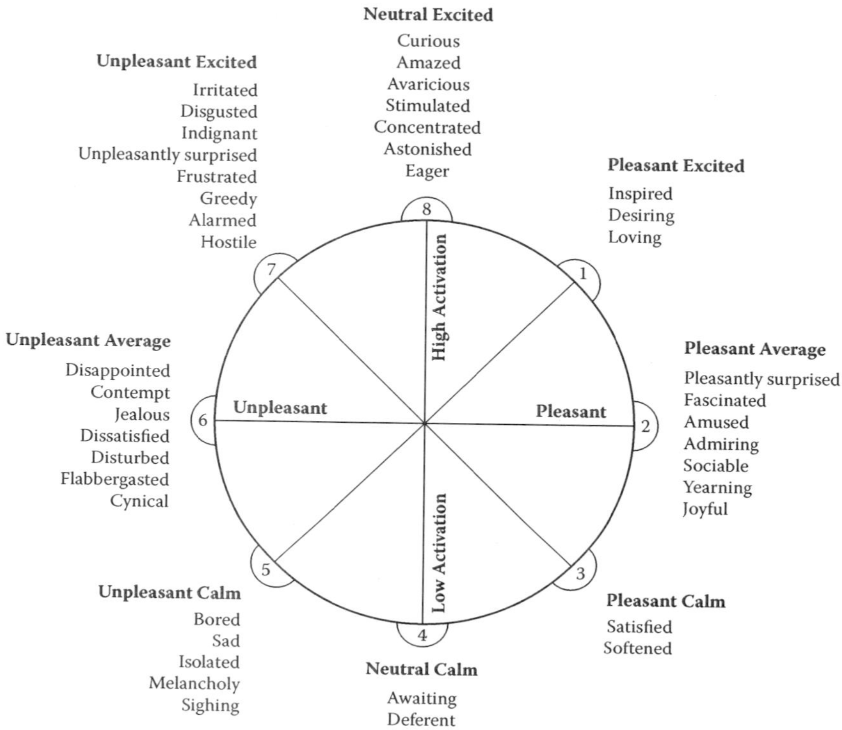

FIGURE 2.2
The circumplex of 41 product emotions developed by Russell (1980) and Watson and Tellgen (185) and adapted by Desmet (2002). (From Desmet, P. M. A. (2002). *Designing Emotions*. Delft, the Netherlands: Delft University of Technology, p. 32.)

Desmet adapted the circumplex of emotions developed by Russell (1980) and Watson and Tellegen (1985) to represent the set of 41 product emotions. The circumplex representation uses the dimensions of pleasantness-unpleasantness and intensity (high-low activation), for the classification of emotions. Larsen et al. (1992) described the circumplex as a circular structure in which single attributes correlate highly with adjacent attributes on the circumference of the circle, correlate less significantly with those attributes one-quarter way around the circle, and correlate inversely with those attributes directly opposite on the circle.

2.1.1.5 Discussion

Among the various theories of emotions, the appraisal theories (relating emotion to appraisal of an emotion-inducing situation) and the transformational theories (relating emotions to person–product relationship) seem to be more relevant in the context of the present study on the emotive quality of products. From the standpoint of the appraisal theories, emotions associated with products can be seen as an outcome of a person's assessment of a particular product. The transformational theories, on the other hand, give an interesting explanation for the strong emotions associated with products like automobiles that convey the person from a logical to a magical world. These theories suggest that product emotions need to be understood in terms of person–product interaction (appraisal) and the desired relationship (transformation) between the person and the product.

2.1.2 Current Approaches for Designing Emotion-Laden Products

We wish to have products that reflect of our own self. Products are no longer bought only for their functional value. In today's scenario, companies focus on the emotional benefits of buying and owning a particular product (Desmet, 2002). Therefore, one of the most important challenges for designers today is to incorporate emotions into products in the early stages of the design process. The importance of designing emotion-laden products can be seen across all cultures and societies.

The research traditions in designing for humans (Norman, 1990, 2004) in general and emotions (Desmet, 2002, 2004, 2008) in particular have encouraged researchers across the world to investigate the role of emotions in design of products. Emotions are now being seen as central to the design process. Products are designed to create an emotional experience and enhanced product attachment. It was therefore important to review the current approaches to designing emotion-laden products. The review focused on various international journal publications as well as conference papers presented at the Design and Emotions Conference at Delft, Netherlands (1999); Loughborough, United Kingdom (2002); Ankara, Turkey (2004); and Gothenburg, Sweden (2006). On the basis of a review of the literature, the current approaches to designing emotions into products could be divided into six categories as Understanding product emotions, Investigating product emotions, Expressing product emotions, Personalizing product emotions, Experiencing product emotions, and Sensing product emotions (Table 2.2).

2.1.2.1 Understanding Product Emotions

> The approach toward understanding emotions focuses on understanding user emotions related to products and differentiation of such emotions.

"Emotion often precedes cognition, although cognition can drive emotions" (Russell et al., 2004). Russell further explained that emotions and cognition cannot be separated and that they work in tandem. Ortony et al. (1988) defined emotions as valence reactions to events, agents, and objects, with their nature being determined by the way in which the eliciting situation is construed. Thus, studies in the field of product emotions should focus

TABLE 2.2

Current Approaches for Designing Emotion-Laden Products

Current Approaches for Designing Emotion-Laden Products	Researchers
Understanding product emotions	Chibber, Porter, Healey and Porter (2004); Desmet (2002); Egmond, Desmet, and Van Der Helm (2004); Enquist (2005); Forlizzi, Mutlu, and DiSalvo (2004); Hanington (2004); Hilton (2004); Hummels (2004); Ortony, Clore, and Collins (1988); Porter, Chibber, and Porter (2002, 2003); Ribas, Battaiola, Montenegro, Andrade, and Queiroz (2004); Savas (2004)
Investigating product emotions	Aslam (2006); Boradkar (2004); Chang and Wu (2004); Collopy (2000); Crilly, Moultyrie, and Clarkson (2004); Desmet (2002); Karana (2004); Lang (1980); Mehrabian and Russel (1977); Pham (1999); Sevener and Asatekin (2004); Tsai, Hung, and Hung (2007)
Expressing product emotions	Bahauddin and Abdullah (2004); Ben-Peshat (2004); Ekman and Friesen (1978); Ekman, Friesen, and Ellsworth (1982); Kaiser and Wehrle (2001); Mura (2008)
Personalizing product emotions	Edwin and Chee (2006); Govers and Mugge (2004); Govers, Hekkert, and Schoormans (2004); Hassenzahl et al. (2000); Hollbrook and O'Shaughnessy (1984); Ingram and Annable (2004); Janlert and Stolterman (1997); Karlsson and Wikstrom (2009); Kim and Moon (1998); Krippendorff (2006); Krippendorff and Butter (1984); Mano (1996); Zimmerman and Ayoob (2004)
Experiencing product emotions	Dewey (1934); Gomez, Popovic, and Bucolo (2004); Multlu and Forlizzi (2004); Henderson, Williams, Castelhano, and Falk (2003); Oustrup and Jeanes (2005); Vuori, Olkkonen, Polonen, Siren, and Hakkinen (2004)
Sensing product emotions	Aydmh, Eren, Erkok, and Sonmez (2004); Schifferstein (2004)

on understanding the role of emotions as valence reactions to objects and exploring the appraisal process that leads to an emotion. Researchers, in their pursuit of understanding how products contribute to emotional aspects of human experience, have proposed different frameworks for product emotions. Egmond, Desmet, and Van Der Helm (2004) reported that in emotion research, a distinction is often made between basic emotions and cognitive emotions. Forlizzi, Mutlu, and DiSalvo (2004) explained another approach toward emotion differentiation by categorizing two levels of emotional expression: short and reflexive emotions and sustained and reflective moods.

Different people can perceive emotions in different ways. This aspect of emotion was discussed by Savas (2004) in the context of social differentiation. He collected empirical data to suggest that social classes also differ in their emotional relationship with products. It is for this reason that many researchers have supported the development of an emotion resource for understanding users' emotions related to products. Chhibber, Porter, Healey, and Porter

(2004) suggested a "pleasure resource" that can be used in research to understand the emotional needs of specific markets. Porter, Chhibber, and Porter (2002), in one of their studies, collected data concerning people's attitude toward products and classified it into a "four-pleasure" framework: physio-pleasure, socio-pleasure, psycho-pleasure, and ideo-pleasure. In the project titled "Real People," Porter, Chhibber, and Porter (2003) documented the various pleasure needs of the user in the form of a DVD resource. The database provides useful information about the user describing his personality, product choice, style choice, and brand choice (Figure 2.3). By documenting the emotive needs and preferences of the users, the developed pleasure resource becomes a useful input to design products that evoke positive reactions.

In addition to developing an emotion resource, researchers have suggested the need to have broader interpretations of pleasure and redefining pleasure related to products (Hanington, 2004). Hilton (2004) discussed various "Attraction Considerations" related to desirability, in the context of judgment of concept attractiveness. The product and emotion navigator (Desmet 2002) is also an image resource that relates the emotions and concerns of users while experiencing various products. Desmet's research aimed to understand the appraisal process that leads to a particular emotion. Other techniques such as interaction and assessment cards (Hummels, 2004), reference boards (Ribas, Battaiola, Montenegro, Andrade, and Queiroz, 2004), and image visualization tools (Enquist, 2005) help in inspiring designers or users to understand their emotions through images.

2.1.2.2 Investigating Product Emotions

> The approach toward investigating emotions focuses on understanding the emotional responses to specific visual characteristics of products and the role of product's visual characteristics in arousing emotions.

Visual stimulation derived from products leads to human–product interaction. Cheskin and Masten Inc. (1987, as cited in Aslam, 2006) reported that imagery is the vehicle that generates interest by the way of "sensation transference" among the target population. According to Crilly, Moultrie, and Clarkson (2004), emotional responses to products can be better understood by considering the cognitive (aesthetic, semantic, and symbolic) responses related to products. Pham (1999) stated that though a designed product can trigger definite aesthetic responses in observers, it is not easy to relate these responses to the characteristics of the product. Therefore, while designing emotions into products, it is useful to investigate the role played by various aesthetic properties (such as color, form, material, graphics, etc.) in stimulating emotional responses. In an experimental research oriented toward a similar viewpoint, Sevener and Asatekin (2004) found that the aesthetic property of "form" is the most dominant factor in product perception. Many researchers have studied the psychological responses to product visual

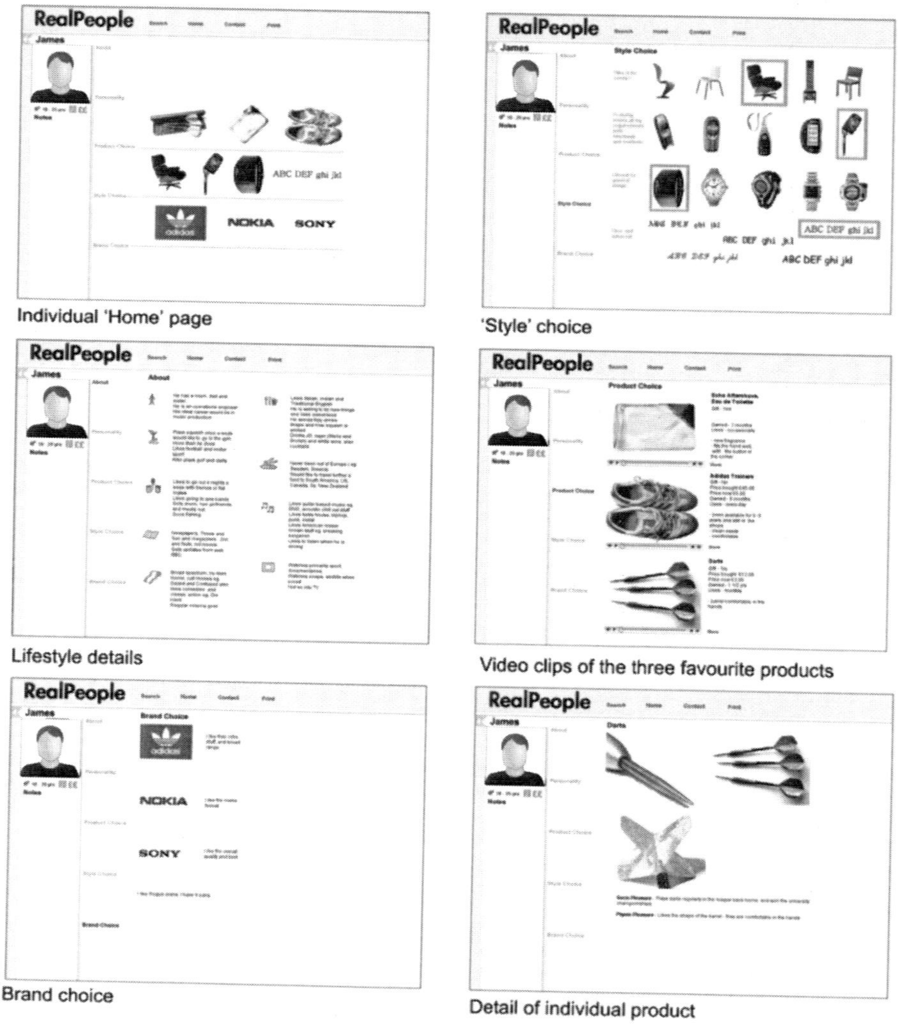

FIGURE 2.3
"Real People": Project for development of pleasure resource. (From Porter, C. S., Chhibber, S., and Porter, J. M. (2003). The development of an inclusive design resource. *Proceedings of INCLUDE 2003: Inclusive Design for Society and Business* [CD-ROM], March 25–28, 2003 (8–10). London: Royal College of Art.)

forms. Chang and Wu (2004) categorized six types of pleasurable forms: bios form, cultural form, aesthetic form, decorative form, unique form, and ideal form. Boradkar (2004) also discussed various types of object skins—rigid, translucent, permeable, elastic, and flexible—in investigation of skins as boundaries.

Similarly, the influence of other visual characteristics such as colors on consumers' perceptions and behavior has also been established in various scientific studies. Aslam (2006) found that color induces moods and emotions. Collopy (2000) found that color is one of the strongest stimuli that our brain receives from the outside world. Aslam (2006) investigated the role of color perception in evoking strong product associations and category imageries. Tsai, Hung, and Hung (2007) established that the product's color rather than its form dominates the overall image perception of a product. In addition to investigating the influence of various properties (such as form and color) on stimulating emotional responses, researchers have also suggested the need to study the relationship and associations among the different aesthetic properties (Karana, 2004). Kumar (2002) suggested an approach of product modeling for aesthetics at early stages of product development to ensure the desired product impression.

The PrEmo (product and emotion measurement instrument) developed by Desmet (2002) is a significant contribution toward investigating the user's emotional responses to specific product characteristics. The 14 puppets on the PrEmo screen (Figure 2.4) represent seven pleasant and seven unpleasant emotions. Each emotion has a three-point scale that is used by participants to report their emotion. Figure 2.4 shows the correspondence analysis of various car designs based on the emotions reported by the participants. Other verbal self-reporting instruments such as SAM (Self-Assessment Manikin) (Lang, 1980), PAD dimensions (Mehrabian and Russel, 1977), and many others have been traditionally used by researchers to measure subjective feelings. Kansei engineering methodology also relates product features to the emotions that they evoke.

2.1.2.3 Expressing Product Emotions

> The approach toward expressing emotions deals with exploring the various ways in which emotional responses can be expressed.

Current approaches toward expressing emotions can be broadly categorized into the study of individual expressions and cultural expressions. Thus, emotional responses have to be studied in the individual space and the cultural space. The latter is often influenced by theories on design and memory. Cultural artifacts often reflect the life philosophy, traditions, and beliefs of its people. These artifacts are like expressions of a culture. This aspect of expressions is also relevant in product design because it gives insights into the emotional influences on the life of people in a specific culture.

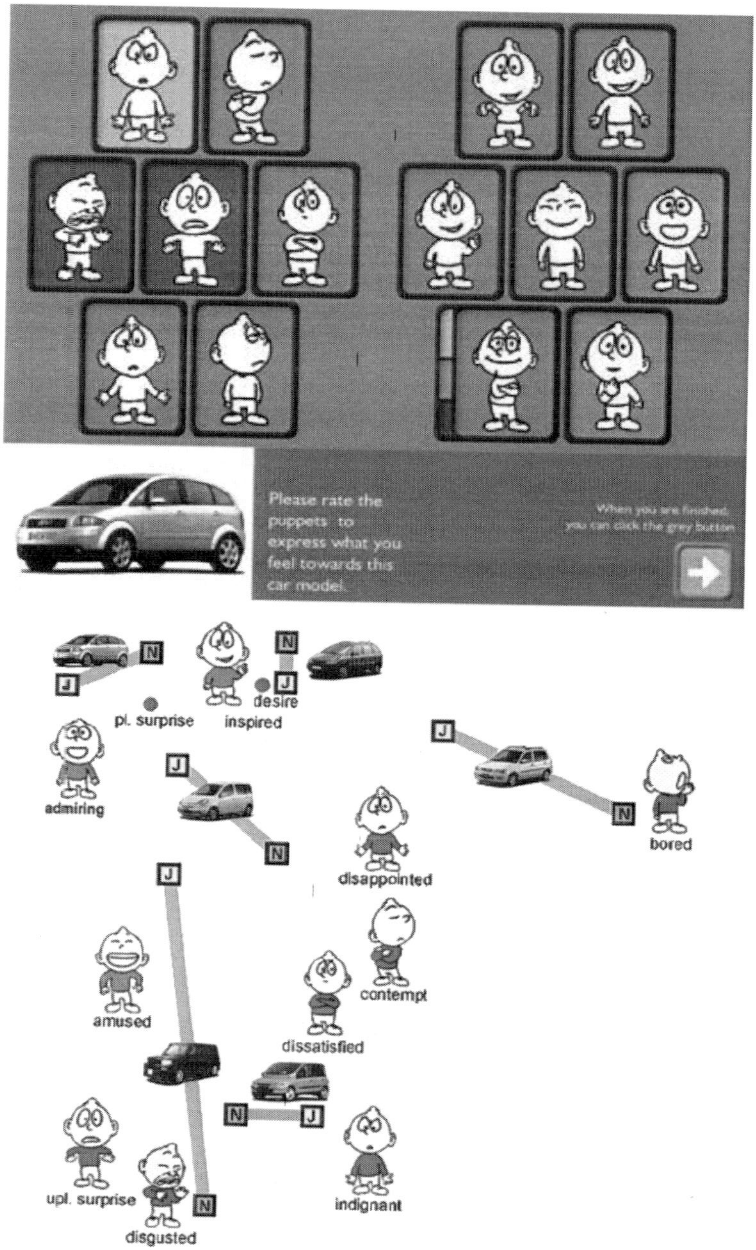

FIGURE 2.4
PrEmo—Product and emotion measurement instrument. (From http://studiolab.io.tudelft.nl/desmet/premo.)

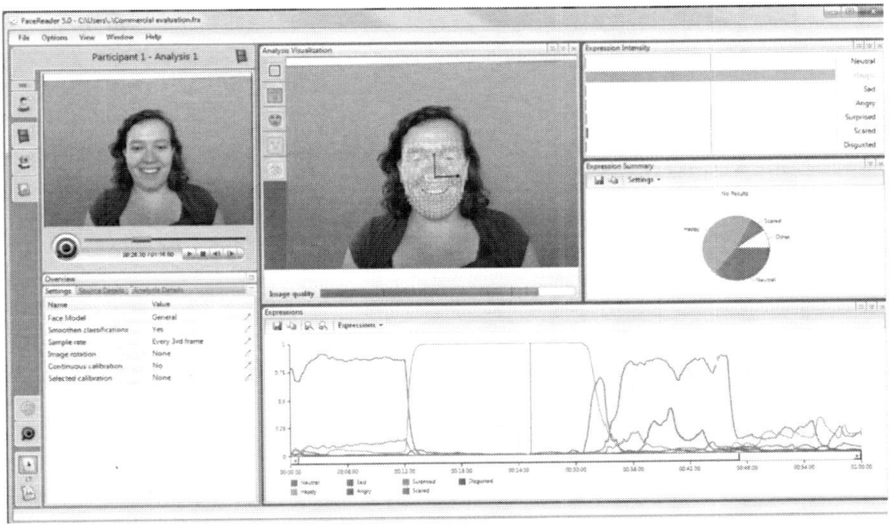

FIGURE 2.5
Noldus Face Reader software for measuring facial expressions. (From Hans Theuws of Noldus Information Technology bv, personal communication, Jan. 21, 2013.)

The individual emotional expressions studied are generally facial expressions and vocal expressions. According to Mura (2008), people mostly communicate their emotions by bodily reactions such as facial expressions, gestures, and body movements. Paul Ekman has been a pioneer in the study of emotions in relation to facial expressions. He listed a set of seven emotions that can be universally expressed on the face such as anger, disgust, fear, joy, sadness, and surprise (Ekman, Friesen, and Ellsworth, 1982). Many researchers (such as Kaiser and Wehrle, 2001) have developed tools for analyzing facial expressions based on the Facial Action Coding system (Ekman and Friesen, 1978). Figure 2.5 shows the Noldus FaceReader software, which is capable of automatically analyzing facial expressions, providing an objective assessment of a person's emotion (http://www. Noldus.com/).

2.1.2.4 Personalization of Product Emotions

> This approach of personalization of product emotions investigates the relationship between the person and the product in enhancing product attachment.

The personality of a person reflects of his overall emotional self and therefore influences his emotional reactions to any situation. Similar to human personality, people perceive expressive or personality characteristics of products (Govers, Hekkert, and Schoormans, 2004), which help us understand and relate to them (Janlert and Stolterman, 1997). Krippendorff (2006) added to this viewpoint, that it is not the physical quality of objects but what they mean to us that determines how we see and act in relation to them. Govers and Mugge (2004, as cited in Desmet, Nicolas, and Schoormans, 2008) explained that the product personality cannot be reduced to distinct product attributes but rather refers to a holistic description of the total product. According to Edwin (2006), the perceived personality characteristics of a product contribute to the final makeup of a product, giving the product its own character.

Person–product interaction leads to an experience. Through these interactions, certain products are able to generate an emotional bond with consumers. This leads to product attachment. Enhancing product attachment is one of the most important strategies for many companies. This attachment can be influenced by visual properties of the product (color, form, material, graphics, etc.) or any other characteristic of a product that can trigger sensory modalities. One way to enhance attachment toward a product is to personalize a product's appearance. In doing so, a person invests energy in the product, and the product is used to express his/her self (Mugge et al., 2004). Personalization of products is about making the product congruent to one's desired experience. Govers et al. (2004) found that congruity between the personality of a person and the personality of his/her product leads to stronger product attachment.

However, this self-expression of a product personality may not always be intentional. A product can have a perceived personality different from what was intended by the designer. Ingram and Annable (2004) explained that a consumer can perceive anthropomorphic and zoomorphic references in a product even without any intentional use of human and animal visual references while designing. These perceptions of product personality can have positive and negative connotations for different user groups. Thus, the personality of a product, like emotions, is personal to the subject (associated user). Just as different people experience different emotions in the same scenario, similarly different people can perceive different personalities in the same product. This concept of product personality resembles celebrity personality. Like products, celebrities also generate emotional reactions from consumers. According to Zimmerman and Ayoob (2004), celebrity products also enjoy strong product attachments and strengthen the consumer–celebrity relationships. Studies like these can provide inputs to designers for the design of products that form similar levels of attachments with consumers.

Krippendorff and Butter (1984) defined product semantics as the study of the symbolic qualities of man-made forms in the context of their use, and the application of this knowledge to industrial design practice. A review of the literature clearly shows different semantic scales and checklists developed by researchers to measure the semantic qualities of products. Kim and Moon (1998) developed 40 semantic scales ranging from awkward to witty. Hassenzahl, Platz, Burmester, and Lehner (2000) suggested and tested various ergonomic factors derived from usability such as comprehensive-incomprehensive, familiar-strange, supporting-obstructing, and hedonic factors such as aesthetic-unaesthetic, inviting-rejecting, or attractive-unattractive for

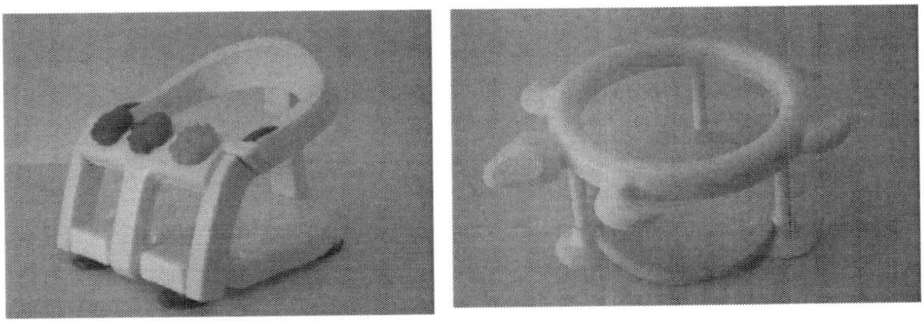

Ratings of product expressions (median values, n=24). The scale ranged from 0=not at all to 100=maximum.

Property	Product A	B	C	D	E
Active	**85**	61	27	50	52
Functional	67	67	74	59	**81**
Durable	**81**	**76**	**88**	50	**79**
Clinical	18	40	**86**	56	44
Amusing	**80**	56	21	20	50
Soft	57	59	76	75	**81**
Unpleasant	50	50	54	50	22
Frightening	21	24	50	12	22
Safe	**81**	56	64	50	60
Restful	38	41	**82**	**78**	**82**

FIGURE 2.6
Product Semantic Analysis (PSA): Products A and B. Product A (left) is a product that was documented to have been involved in drowning accidents. (From Karlsson, M.A., and Wikstrom, L. (2009). Safety semantics. A study on the effect of product expression on user safety behaviour. *Contemporary Ergonomics Society Annual Conference* 1984–2008, pp. 609–613.)

describing the emotive quality of a product. Hollbrook and O'Shaughnessy (1984) developed a tool to measure emotions in advertising on seven dimensions of emotional responses: entertainment, confusion, relevance, favorability, empathy, familiarity, and alienation. The adjective checklist developed by Mano (1996) addresses the sensory-aesthetic aspect of emotional experiences and relates them to pleasure and the arousal dimension of emotion. Figure 2.6 shows the Product Semantic Analysis tool developed by Karlsson and Wikstrom (2009) to measure the semantic quality of a product. The figure shows the results of an assessment of an electric cooker by customers and designers using the Product Semantic Analysis Tool.

2.1.2.5 Experiencing Product Emotions

> This approach of experiencing emotions deals with the overall experience associated with a person–product interaction.

While creating new products, designers create new experiences. Design processes always involve a consumer and a product. The emphasis on human experiences and person–product interaction in design research is growing among designers and design researchers across the world.

According to Dewey (1934), the object of experience contributes to the quality of experience with its intrinsic material to shape the experience in an intellectual, emotional, or practical form. He also suggested that any experience has two major components: the object being experienced and the experience itself. Multlu and Forlizzi (2004) described the emergence of emotions as a part of a deterministic system of experiences. The experience during a product–person interaction is often influenced by ambient conditions. Therefore, while studying the experience during any event, it is important to understand the collective experience that is generated. Gomez, Popovic, and Bucolo (2004) explained that the context and situation have a dominant impact on the human experience. Gomez et al. (2004) studied the relationship between driving experience and traffic conditions. They found that negative feelings are magnified if usability problems occur in high-traffic conditions. Thus, the overall experience becomes negative owing to these associations. Hence, one cannot treat experience with an object in isolation without understanding the effect of other conditions in the situation under which the interaction (human–product) occurs.

Designers are constantly trying to create new customer experiences. But since experiences are personal in character, they can vary with individuals. This makes it difficult to design desired customer experiences. Figure 2.7 shows the EED-Elemental experience design technique developed by Mette Kristine Oustrup for creating great customer and brand experiences. This technique combines the Western focus on emotions and personality with the Asian art of architecture and elemental design (Oustrup and Jeanes, 2005).

EED categorizes people, products, ads, services, stores, and experiences on one of the five levels: Earth, Water, Wood, Fire, and Metal–Air. Each of these elements has specific characteristics. People can be categorized by these elements on the basis of their personalities, behavior, and preferences, while products and spaces are assigned to these elements on the basis of their aesthetic characteristics. The experiences associated with different elements are characterized on the basis of related emotions.

According to Oustrup and Jeanes (2005), EED allows designers to plan, measure, and design experiences. Unlike the traditional research techniques based on categorizing people on personality types, EED determines elemental types that are more visual and inspirational in the creative process. EED also reduces the number of questions to be included in a survey for determining the elemental type as compared to long questionnaires used for determining traditional personality types.

2.1.2.6 Sensing Product Emotions

This approach of sensing emotions deals with investigating the role of different sensory modalities in triggering product emotions.

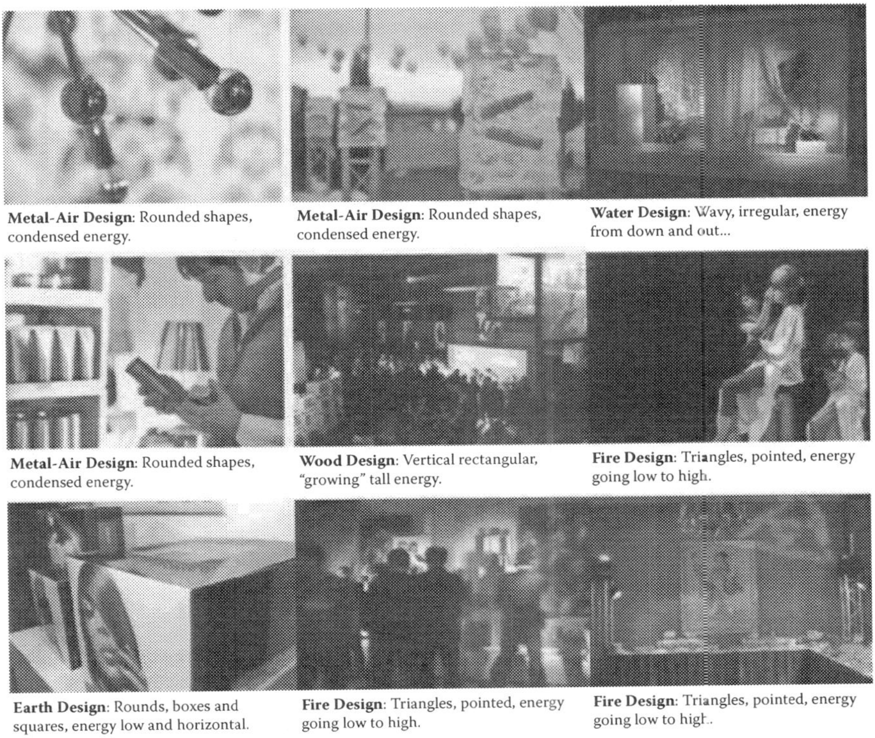

FIGURE 2.7
EED-Elemental experience design technique. (From Oustrup, M. K., and Jeanes, M. (2005). MOOD consumption theory: A human-focused marketing tool. Presented at ESOMAR, Annual Congress, September 18–21, 2005 (6), Cannes.)

The five senses of the human body represent the sensory modalities. We are able to perceive products through sensory messages that are received by these modalities during person–product interactions. According to Schifferstein (2004), the same product can evoke different sensory messages because some sensations trigger feelings and emotions; some evoke memories or associations with other objects. Schifferstein conducted single-modal, cross-modal, and multimodal experiments to understand the roles of individual modalities and the correspondence between modalities in product experiences. It is possible to evoke positive or negative product experiences by considering the role of sensory modalities of the user. Schifferstein in his multimodal approach for generating product experience found how inputs from various sensory modalities are integrated into a product experience. This concept of sensory modalities and bodily experience can be used for designing for self. Aydmh, Eren, Erkok, and Sonmez (2004) investigated, through their experiments on bodily experience, whether experiencing personally, using the "self" as a means

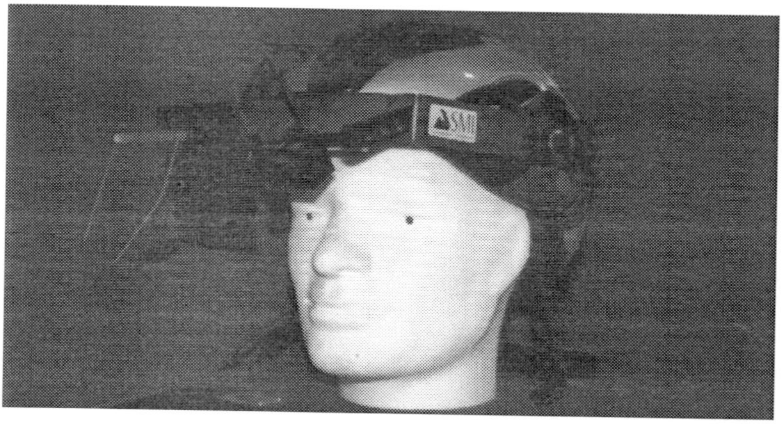

FIGURE 2.8
Head-mounted eye-tracking device developed by SensoMotoric Instruments.

of experience, can be a supportive method to create potential experiences. It is seen that designers often use various body storming techniques in which they immerse themselves into the user experience for which design improvement is desired.

The sensory responses can be studied with the use of instruments that measure heart rate, skin conductance, eye movement, and brain waves. Figure 2.8 shows a head-mounted eye-tracking instrument used to measure the pupil fixation and study the gaze path during visual interaction. The gaze behavior reveals the variability in the emotional perception of the user toward the stimuli under investigation. Researchers (Vuori et al., 2004); Henderson et al., 2003) have used this methodology for explaining individual differences in image perception.

2.1.2.7 Discussion

The review of the literature leads to an understanding of the extensive research being undertaken by researchers in exploring the role of emotions in designing across product categories. It was found that though "the emotional concerns of users towards products" (Desmet, 2002; Ekman et al., 1982; Forlizzi et al., 2004; Hanington, 2004; Hilton, 2004; Kaiser and Wehrle, 2001; Multlu et al., 2004; Mura, 2008; Schifferstein, 2004; Lang, 1980) and "the influence of products on triggering emotions" (Boradkar, 2004; Chang and Wu, 2004; Crilly et al., 2004; Sevener and Asatekin, 2004; Oustrup et al., 2005) have been studied independently by researchers in the past, yet no appropriate research framework could be identified that could be used to investigate both these aspects in the context of the present study of motorbikes in India.

2.2 Theoretical Applicability of the Research Hypothesis

2.2.1 Context-Specific Segmentation of Product Ownership

2.2.1.1 Market Segmentation

Wendell Smith (1956), one of the originators of segmentation thinking, defined product differentiation as being concerned with the bending of demand to the will of supply. In times of mass production, product differentiation is an effective strategy to gain competitive advantage. The basic principle of market segmentation is that markets are not homogeneous, and therefore it is important to differentiate product offerings for different customer groups. According to Greengrove (2002), not all customers are the same and markets consist of a number of "segments," each consisting of "homogeneous" customers. According to Kara and Kaynak (1997), market segmentation aims to divide markets comprising individuals into groups whose characteristics are relatively homogeneous within each set or segment and heterogeneous between segments, based on an identified set of variables.

Market segmentation is a widely adopted concept among marketing managers across the world, to gain a holistic understanding of a particular market, to predict consumer behavior and to identify new market opportunities for commercial benefit (Heok, Gendall, and Esslemont, 1996). McDonald (1999) defined a market segment as a group of customers or consumers with similar needs. McDonald and Dunbar (1995) defined market segmentation as the process of dividing customers into different segments on the basis of similarity in their characteristics and similar needs.

2.2.1.2 Segmentation Approaches

Greengrove (2002) noted two main segmentation approaches: needs-based and characteristics-based. The needs-based segmentation (McDonald, 1999) approach is based on segmenting the market on the basis of end user needs, whereas the characteristics-based segmentation approach (McDonald and Dunbar, 1995) is based on segmenting the market on the basis of the characteristics, attitudes, or behavior of customers. Fuller, Hanlan, and Wilde (2005) discussed a priori and post hoc (Dolnicar, 2004; Kara and Kaynak, 1997; Wind, 1978) approaches to segmentation. In a priori segmentation, customers are assigned to segments designated a priori on the basis of the chosen variables, whereas, in post hoc segmentation, customers are clustered on the basis of the chosen variables to produce segments post hoc (Wind, 1978). Though the a priori approach may generate segments that are internally consistent in chosen variables, this does not necessarily mean that all segment members will respond in the same way to marketing stimuli (Hoek et al., 1996). On the other hand, a post hoc approach may result in segments with similar attitudes and behavior traits; however, this does not necessarily mean that all segments will be internally consistent in terms of the demographic characteristics (Hoek et al., 1996).

On the basis of the foregoing segmentation approaches, researchers and marketing practitioners have identified various variables for segmenting markets. Markets are commonly segmented on various variables such as geographical boundaries, demographic characteristics, consumer lifestyle, values, attitudes, and usage patterns.

2.2.1.3 Need for Context-Specific Segmentation of Product Ownership

The growing personalization of consumer behavior patterns is not well explained by the earlier approaches to segmentation, including socio-demographic and economic criteria alone (Blattberg, Peacock, and Sen, 1976; Gonzalez and Bello, 2002). There is a growing interest in alternate segmentation approaches seeking to measure less tangible consumer characteristics such as lifestyle, personality, image, and benefits (Alpert, 1972; Myers, 1976; Sirgy, 1982). Rather than demography, it is the difference in buyer attitudes, motivations, values, patterns of usage, aesthetic preferences, and degree of susceptibility that have become central to marketing planning (Yankelovich, 1964). Yankelovich (1964) emphasized the importance of finding new methods for segmentation analysis such as segmenting of markets according to consumer values and style preferences. Today, the benefit-based segmentation approach has become the preferred technique for successful product positioning, new product introduction, pricing, and advertising (Botchen, Thelen, and Pieters, 1999). The benefit segmentation approach is based on measuring the consumer value systems and consumer thoughts about various brands in the product category of interest (Haley, 1968).

Despite the ongoing proliferation of a number of variables for market segmentation, it is extremely difficult to find a single approach that may be appropriate for segmenting different markets. It is generally recognized that there is no single solution, as each approach has certain merits and limitations, depending on the products and markets being considered (Kara et al., 1997; Wind, 1978). Fuller et al. (2005) suggested that effective market segmentation is context specific and allows for creative approaches, depending on the research problem. The literature clearly indicates a trend toward identifying new context-specific approaches to market segmentation. A context-specific approach is likely to be based on causal relationships between consumers and products as compared to other descriptive methods of segmentation.

2.2.1.4 Social-Cultural Segmentation of Biking in India (as Presented in Chapter 1)

The first chapter of the book presents a theoretical model of six biking segments: bread-and-butter biking, travel-and-comfort biking, social-networking biking, dexterous and exploratory biking, high-esteem biking, and religious biking. The model was developed to represent the social-cultural segmentation of biking in India. This segmentation was based on studies conducted across various cities in India and reflects the social reality of biking in India. It was observed during studies that biking embodies the relationship between the biker and the bike and is influenced by bikers' emotive needs for purchasing, riding, or owning a motorbike. The heterogeneous needs of biker groups have given rise to distinct biking segments. The differences in emotive needs of biking were reflected in the biker's lifestyles, attitudes, values, usage patterns, and the benefits that they sought from riding, purchasing, or owning a motorbike. The proposed segmentation can be seen to be similar to the benefit segmentation approach discussed above. Similar to the benefit segmentation approach, grouping bikers on the basis of their emotive needs lead to multiple segments each with a set of exclusive and some common emotive needs. It is, however, the total configuration of emotive needs that differentiated one segment from another.

The literature on market segmentation, as reviewed above, supports the applicability of the social-cultural segmentation of biking in India, posited in Chapter I. But the social-cultural segmentation of biking can be ascertained only by investigating and evaluating the total configuration of emotive needs associated with each biking segment. A survey-based methodology was therefore used by the writer to empirically test the efficacy of the proposed segmentations.

2.2.2 Directions for Investigating the Emotive Quality of Product Ownership

The previous section investigates the theoretical applicability of the social cultural segmentation of biking in India. The research premise, presented in the first chapter, suggests a possible difference in the emotive needs and the emotive quality of the identified biking segments. During the two exploratory studies, it was observed that in social reality a motorbike cannot be seen in isolation from its biker. A bike and biker together complete a biking or motorbike ownership experience. Thus, it was believed that the emotive quality of a biking segment should be studied by evaluating the bike's image or personality in relation to the self-image or personality of the biker. The review of relevant literature threw up ways of investigating the emotive quality of product ownership experiences in the context of the present study.

2.2.2.1 Psychological Ownership

Harris and Young (1983) explained that the art of selling involves bringing together the particular item that meets a particular buyer's need. According to Harris and Young (1983), most consumer products have a number of aspects that can appeal to prospective buyers such as the investment value of the purchase, the economical operation of the product, or the pure emotional appeal of product ownership.

Research in a wide variety of fields emphasizes the importance of possessions and ownership in influencing human attitudes, motivation, and behavior (Dyne and Pierce, 2004). According to Furby (1978), a sense of possession is the core of psychological ownership. Possessive feelings are ubiquitous, can refer to tangible or intangible objects (Beaglehole, 1932; James, 1890), and can occur based on legal ownership or in the absence of legal ownership (Wilpert, 1991). Researchers have reported a close connection between possessions, feelings of possession, and feelings of ownership (Dyne and Pierce, 2004). It is often seen that people tend to equate feelings of possession with feelings of ownership (Dittmar, 1992; Furby, 1978).

This book proposes a relationship between the needs of product ownership and buyer's choice of product. In this context, it is important to understand the link between feelings of ownership and self-concept of the owner.

2.2.2.2 Self-Concept and Self-Congruity

Consumers have evaluations of themselves that contribute to their self-image (Grubb and Hupp 1968). Grubb and Grathwohl (1967, p. 24) defined self-concept as "the self is what one is aware of, one's attitudes, feelings, perceptions, and evaluations of one's self as an object." De Chernatony and de Mello (1995) stated that self-concept theory has a multidimensional nature. Self-concept consists of four components: actual self-concept, ideal self-concept, social self-concept, and ideal social self-concept (Schiffman, Bednall, Cowley, O'Cass, Watson, and Kanuk, 2001).

Researchers agree that one's self-concept is extremely valuable to an individual and must be protected and enhanced (Belk 1988; Grubb and Grathwohl, 1967). Also, people like to maintain a consistent (self-consistency motive) and positive (self-enhancement motive) view of one's self (Mugge, Schifferstein, and Schoormans, 2006). According to Sirgy (1982), products are a medium by which an individual can symbolically define and express his/her unique self. People like to acquire products that reflect their self-image (Grubb et al., 1967; Kleine, Kleine, and Kernan 1993; Osman 1993), or help in communicating their self-image to others (Richins 1994a, 1994b). People's tendency to behave consistently with their self-image has led to the concept of self-congruity. Self-congruity is a theoretical framework that predicts that a consumer will exhibit a preference for the brand or product whose image most closely corresponds to his or her own self-concept (Barone, Shimp, and Sprott, 1999).

Thus, anchoring the product or brand personality to the user personality can lead to effective market segmentation (Webb and Gountas, 2006). However, literature emphasizes that more research may be required to study real-life situations to fully understand the congruity relationships between person and the product (Mugge et al., 2006). This book drew inspiration from these points and attempted an effective segmentation of motorbikes within the social cultural precincts of the Indian two-wheeler market.

The understanding gained from the review of literature and the findings of the two exploratory studies, presented in the first chapter, were correlated to arrive at the following two significant directions for investigating the emotive quality of products.

2.2.2.3 Person–Product Personality Gives an Emotive Quality to Product Ownership Experience

The literature on product personality (Edwin, 2006; Govers et al., 2004, Janlert and Stolterman, 1997; Krippendorf, 2006; Mugge, 2007) supports the viewpoint that similar to human personality, people perceive personality characteristics of products, which helps them understand and relate to those products. But just as different people can experience diverse emotions with the same product, similarly, the product's personality can also vary with individuals. The emotive quality of a product ownership experience therefore cannot be seen in isolation from the owner or the user of that product. The person–product personalities should be evaluated together on similar attributes to assess the evolved personality of the related product ownership experience. This conjoint personality indicates the product's emotive ability to satisfy or dissatisfy a person's emotive needs and desires. It is this person–product personality that gives an overall emotive quality to a product experience.

Though many studies have been cited that emphasize the importance of person–product interaction in shaping the overall experience (Dewey, 1934; Gomez et al., 2004; Multlu et al., 2004) and relationship between the personality of the person and the product in enhancing product attachment (Govers et al., 2004; Mugge et al., 2006), yet no significant study could be identified that delineates the person–product personality in defining emotive qualities for specific product ownership experiences.

Thus, in the present study, the emotive quality of different biking segments, discussed earlier, is studied by evaluating the biker–bike personality associated with each biking segment.

2.2.2.4 Emotive Words Describe the Emotive Quality of Product Ownership Experience

"Humans are affective beings, motivated to action by a complex system of emotions, drives, needs and emotional conditioning in addition to cognitive factors" (Claire Dormann, 1984). There has been extensive research in the field of emotions by different researchers. The attempt has generally been to study the mechanisms that develop emotions. In doing so, different researchers such as Arnold (1996), Ekman et al. (1982), Frijda (1987), Gray (1982), Izard (1972), McDougall (1926), Oatley et al. (1987), Plutchik (1980), Tomkins (1984), Watson (1930), Weiner and Graham (1984) have proposed several ways of differentiating emotions.

Since every emotion may not be relevant to product experience, Desmet (2002) identified a set of 41 emotions relevant to the study of products. While Desmet's emotion words define the emotional end state of a user experiencing a product, the propensity of bikers to associate with these emotion words is something that needs to be investigated. Also, most of the emotion words like *amused, satisfied, awaiting, deferent, bored, isolated,* and *irritated* do not define the emotive quality or personality of a motorbike. For example, we do not talk of a Royal Enfield Bullet as an "amused bike" or a "satisfied bike." Instead, we know it as a "sturdy bike" or a "heavy bike." The personality of a product (bike) can be defined by emotive words instead of emotion words. The emotive words have a tendency to trigger emotions. The emotive words are, therefore, productive of or directed to emotions.

One such group of emotive words is *adjectives*. Adjectives in a language help in evoking emotions by defining an object. Mano (1996) developed adjective checklists to measure the sensory-aesthetic aspect of emotional experiences and relate them to pleasure and arousal dimension of emotion. Many researchers (Hassenzahl et al., 2000; Hollbrook et al., 1984; Kim and Moon, 1998) have developed semantic scales and set of emotive words as tools to measure emotions in product design and advertising. More recent studies have generally used, as suggested by Sirgy (1982), only those adjectives that are most related to the products being tested.

According to Goldman (1995), "critics and laypersons invariably ascribe properties to artworks which are beyond the ones described in purely physical terms" (p. 17). He classified the various words that are used for ascribing these properties to artworks into eight evaluative aesthetic terms: evaluative terms (beautiful, ugly, sublime, dreary), formal terms (balanced, graceful, concise, loosely woven), emotion terms (sad, angry, joyful, serene), evocative terms (powerful, stirring, amusing, hilarious, boring), behavioral terms (sluggish, bouncy, jaunty), representational terms (realistic, distorted, true to life, erroneous), second-order perceptual terms (vivid, dull, muted, steely, mellow), and historical terms (derivative, original, daring, bold, conservative). According to him, similar to artworks, products like motorbikes are often described in formal terms ("a precision machine"), evocative terms ("a powerful engine"), behavioral terms ("a confident posture"), representational terms ("a naked motorbike"), perceptual terms ("vivid colors"), historical terms ("a traditional model"), broadly evaluative terms ("a beautiful scooter"), and emotion terms ("a joyful ride"). Goldman's evaluative aesthetic terms were found to be relevant for the classification of emotive words related to motorbikes in the context of the present study.

2.2.3 Research Objectives

The primary aim of this book was to develop an emotion-centered research framework for product styling. In order to meet the primary aim, the following research objectives were identified for this study:

1. To verify the social-cultural segmentation of biking in India.

   To investigate the difference in the emotive needs of the bikers belonging to bread-and-butter biking, travel-and-comfort biking, social-networking biking, dexterous and exploratory biking, and high esteem biking segments.
2. To evaluate and compare the emotive quality of different biking segments.

   To develop a biker–bike personality measurement instrument to investigate the biker–bike personalities associated with bread-and-butter biking, travel-and-comfort biking, social-networking biking, dexterous and exploratory biking, and high-esteem biking segments.
3. To formulate an emotion-centered research framework for product styling

   To synthesize the findings of the study into a generic, emotion-centered research framework that focuses primarily on managing emotions in design innovation.

2.3 Significance of the Study for Motorbike Design

2.3.1 Designing Emotion-Laden Motorbikes

The emotion design strategy of major global motorbike manufacturers such as Honda, Yamaha, Harley-Davidson, Ducati, and BMW was examined by the writer, in relation to their choice of emotions and expressions in design.

2.3.1.1 Form Follows Emotion

Emotions play a significant role in the descriptions of motorbikes and biking experiences among manufacturers worldwide. Various motorbike manufacturers across the globe can be clearly differentiated on the basis of their choice of emotions and their subsequent reflection in their designing of bikes.

Joy is the prime emotional expression for the Honda Motor Company. Honda expresses this emotion at all levels by following a unique philosophy of three joys, which are the joy of selling, the joy of buying, and the joy of manufacturing (Honda, 2009). Yamaha Motor Company, on the other hand, believes in its corporate mission of creating *Kando* (Japanese word for simultaneous feeling of deep satisfaction and intense excitement) in people's lives. Yamaha associates emotive keywords like sex appeal, exciting, self-expression, fun, maneuverability, and convenience to define its "art of motorcycling" (Yamaha, 2009). Italian manufacturers such as Ducati define its motorbike styles with emotive words such as aggressive, attractive, inspiring, sleek, and seductive (Ducati, 2009). On the other hand, BMW symbolizes the character of German engineering by creating perfect rider machines to deliver sheer driving pleasure, luxury, comfort, and excitement (BMW, 2009). The mission of enduring American brands such as the Harley-Davidson Motors is to ignite the passion among its riders by creating simple, familiar, ambitious, and above all unforgettable experiences (Harley-Davidson, 2009).

A manufacturer's choice of emotions gives a distinct emotive quality to a particular brand. Motorbike designers give great importance to the aesthetic expression of the brand emotions. The semantic transformation of the brand emotions to aesthetic design features is clearly visible in the product styling and communication strategy of various motorbike manufacturers. The core brand values and emotions are embedded in a bike through the use of various design cues. Motorbikes are immediately recognized and related to their parent brand because of the consistent design philosophy adopted by manufacturers worldwide.

Figure 2.9 shows the images of five different racing sports bikes designed by five global motorbike manufacturers. Though all the bikes shown in Figure 2.9 are essentially sports motorbikes designed for racing, they can be differentiated because of their distinct styling genre. The Yamaha YZF-R6 aims at creating a *Kando* experience through its sharp razor handling and styling. The Yamaha "fly-by-wire" throttle mechanism complements the sharp-styled bodywork in delivering ultimate acceleration and excitement. The "catch the extreme" (Yamaha-YZF-R6, 2009) communication slogan of Yamaha R6 matches its distinct character in style and performance. The Honda CBR1000RR, on the other hand, is more balanced in style as compared to the distinctive Yamaha R6. Honda describes this bike as "the top of superbike food chain" (Honda-CBR1000RR, 2009). Unlike Yamaha, which boasts of its superb acceleration, Honda highlights its combined antilock braking system and proven handling and performance in product communication. The Honda CBR1000RR has a more static, sturdy, and powerful image as compared to the Yamaha R6, which has more dynamic body lines. The overall product value is reflected in all Honda bikes as compared to radical styles and distinctive performance features that Yamaha offers. The styling differences between a Yamaha and a Honda can be seen as the reflection of their emotive design philosophy.

The BMW HP (high performance) sports bikes showcase the might of German engineering. Unlike its Japanese counterparts, the BMW HP-2 exposes most of its mechanical components: the fully adjustable Ohlin's

FIGURE 2.9
Examples of sports bikes of different global motorbike manufacturers.

sports front and rear suspension, radial-mounted Brembo Monobloc brakes, carbon fiber aerodynamic fairing, forged racy wheels, and a Moto GP–inspired cockpit that computes racing data to highlight the "passion for performance" (BMW-HP2 Sport, 2009) image of this bike. The body work too is more mechanical and geometric as compared to the organic forms seen in the Yamaha R6 or Honda CBR1000RR. On the other hand, the look and stance of a Ducati 1098 represents the Ducati heritage. Like all Ducati bikes, this bike can be defined as "pure and simple" (Ducati-1098, 2009). The monochrome body color gives a no-nonsense look to all Ducati bikes. Ducati tries to make bikes that appear racy even when they are standing still. The trademark Ducati features like high tail section, compact front end, under-seat twin silencers, and single-sided swing arm differentiate this Italian beauty from other sports bikes in its category.

The Harley-Davidson sports bikes (Sportster series) take the Harley DNA onto the racetracks. The Harley-Davidson Nightster model shows off its low suspension, bare-knuckle styling, chopped front fork, chopped rear fender, chrome slash-cut exhausts, black mounted headlamp and visor, black low-rise handlebars, peanut fuel tank, and black sliders on front forks with black gaiters. The solo seat of this bike promises an unforgettable Harley experience to its rider. Unlike the feeling of Honda's "Joy" or Yamaha's "Kando," the Nightster promises a "kick to the gut" (Harley-Davidson Nightster, 2009) riding experience.

Design can be used to strategic advantage by enhancing the brand identity of a company (Schmitt and Simonson, 1997; Stompff, 2003; Borja de Morota, 2004). According to Karjalainen (2007), companies must develop products with designs that not only appear attractive but also carry distinctive references to the core values and the character of the brand. Chan and Sanders (2004) emphasized that the brand essence reflects the personality of the brand and allows customers to emotionally connect with the brand. Figure 2.9 illustrates that in addition to informing us about functional properties, motorbike design acts as a carrier of the brand's emotions as well. The symbolic associations and meanings related to design features give a distinct personality to motorbikes associated with a particular brand. Designers incorporate various implicit value-based design cues and explicit artificial design cues (Karjalainen, 2004; Karjalainen and Warell, 2005) while designing new motorbikes. The consistent use of these design cues enables companies to differentiate their products from their competitors.

2.3.1.2 Emotion Follows Form

The previous section explained the differentiation in the manufacturer's choice of emotions and its reflection in the styling of motorbikes. The discussion covers the aspects of emotions from the manufacturers' viewpoint, but it does not discuss the users' emotions associated with different motorbikes. Though a motorbike manufacturer may integrate certain emotions into its corporate or design philosophy, yet these emotions can have different meanings for different users. A particular styling characteristic may not communicate a similar emotive feeling across different bikers. A motorbike that evokes excitement among a particular user may fail to stimulate another person or the same person at a different point of time. This leads to the question: how does the manufacturers' choice of emotions result in a desirable product?

The answer to this question lies in the fact that the manufacturer's choice of emotions, rather than becoming a direction for design, actually represents an emotive framework for design research. The research and development team of all major motorbike manufacturers are constantly studying the motorbikes and bikers within this research framework. For example, the Honda philosophy of "the three joys" would influence its design research strategy for investigating its current and potential users. Similarly, Yamaha,

BMW, Ducati, and Harley-Davidson follow their own research framework for new product development based on their choice of emotions.

The motorbike design process can be broadly divided into six phases: (1) the research phase, (2) 2D design development phase, (3) 3D design development phase, (4) show/validation model development phase, (5) pre-production prototype and development phase, and (6) design implementation phase (Xenophya design, personal communication, July 23, 2009). During various interactions with motorbike designers in India, Japan, Germany, Italy, and the United Kingdom, it was found that most motorbike manufacturers and design consultants divide the design process into similar phases as discussed above. With easy access to modern-day technology, most companies use similar tools, equipments, and machinery during various phases of product development. However, it is the first phase, that is the research phase, that differentiates the product development process of different manufacturers and sets the direction for designing emotion-laden motorbikes. Different companies follow different research methods on the basis of their selected framework of emotions.

The discussion suggests that all motorbike companies follow a closely guarded research methodology that is defined within the boundaries of its choice of emotions. The present study, on the other hand, deals in developing an emotion-centered research framework focusing on the users' emotions within the social cultural environment of India. Figure 2.10 illustrates the significance of the present study with respect to the motorbike design process.

The literature has emphasized the importance of emotions in motorbike design and the significance of the present study for motorbike industry. Also, though many studies have cited examples of the role that emotions

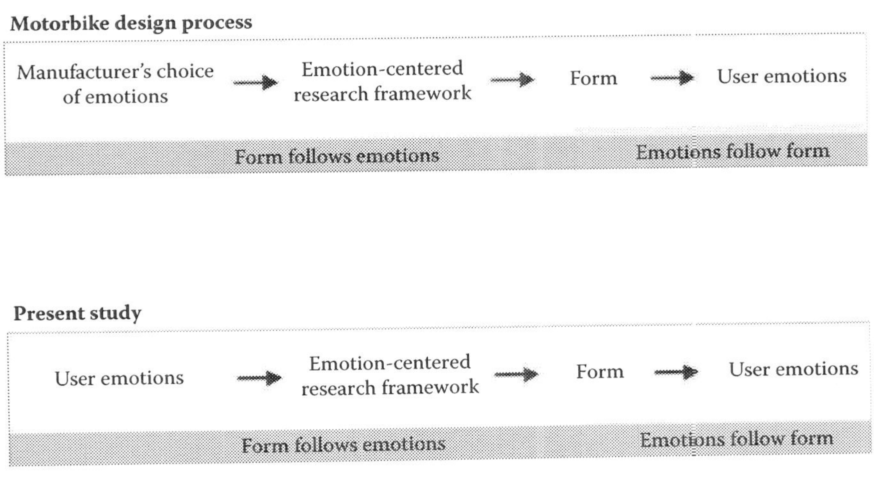

FIGURE 2.10
Significance of the present study with respect to the motorbike design process.

play in designing of motorbikes, they were primarily restricted to the Western culture of biking. No significant study could be cited in the context of motorbikes in India, Southeast Asia, South America, and so forth. This book, therefore, significantly evolves a method of study that includes the development of measurement instrument, sampling of bikers and bikes, survey questionnaire, procedure for data collection, and techniques used for analysis of data.

3

Evolution of Research Methodology

To conduct this study, the writer evolved a method that incorporated the development of a self-reporting measurement instrument, design of survey questionnaire, sampling of bikers and bikes, and selection of procedure for data collection and analysis.

3.1 Development of Biker–Bike Personality Measurement Instrument

The review of the literature, as explained in the previous chapter, gave directions for identifying a cluster of adjectives that characterize motorbikes in India. In this context, an online study was conducted to identify a set of adjectives that induce emotions among motorbike users and manufacturers in India.

An online word-count frequency investigation formed the basis for developing a cluster of adjectives related to motorbikes in India. The Inspyder Insite 2.0 (http://www.inspyder.com/products/) word-count software was used for the purpose of this study. The study was divided into three phases.

3.1.1 Phase 1: Compiling the List of Adjectives in English Language

In the first phase, the *Oxford Advanced Learner's English Dictionary*, 7th ed. (OALD, 2005), database was analyzed to identify the list of adjectives. An extensive database was created in Microsoft Excel 2007, by personally analyzing and tabulating the grammatical usage of each word given in the dictionary. It was found that the dictionary database of 42,329 words comprises 8,794 adjectives, 5,630 verbs, 28,913 nouns, 614 abbreviation, 969 adverbs, 54 conjunctions, 59 determiners, 247 exclamations, 103 pronouns, and 119 prepositions. The identified 8,794 adjectives were used for further study.

3.1.2 Phase 2: Online Word-Count and Frequency Study on Motorbike Websites in India

An online word-count and frequency study was conducted using Inspyder Insite 2.0 software. The online investigation was done on six Indian two-wheeler manufacturers' websites: (1) Hero Honda Motors Limited (http://www.herohonda.com/), (2) Bajaj Auto (http://www.bajajauto.com/), (3) TVS Motors (http://www.tvsmotor.in/), (4) Royal Enfield Motors (http://www.royalenfield.com/), (5) Yamaha Motor Company India (http://www.yamaha-motor-india.com/), and (6) Honda Motor Company India (http://www.honda2wheelersindia.com/), as well as India's largest biking community portal, XBHP (http://www.xbhp.com).

A total of 31, 61,092 (unique words: 90,210) words were extracted from more than 1,470 web pages and 3,855 web links of the seven motorbike websites. The identified 31,61,092 words were cross-linked to the previously identified database of 8,794 adjectives. A total of 4,38,583 adjectives (unique adjectives: 3,643) were identified from the complete database of words. Table 3.1 shows the results of web analysis. The identified 3,643 unique adjectives were subjected to frequency and percentage analysis using SPSS 13.0 software. The objective of the investigation was to identify the unique adjectives frequently used by manufacturers, owners, and intending buyers of bikes in India. The company-wise frequency percentage was calculated for each adjective. Table 3.2 show the list of 1,000 most frequently used adjectives, identified on the basis of the average frequency percentage.

The most frequently used 1,000 adjectives were analyzed in detail to identify a final list of adjectives related to motorbikes. For example, though words such as *motor* and *public* were used frequently as adjectives on the websites under study, yet they were omitted from the final list because they were used mostly used in the form of "motor bikes" and "public limited company." Only those adjectives that were mostly used for describing motorbikes, on the websites under study, were selected for further analysis. Table 3.2 shows the list of adjectives (highlighted) that were selected for further analysis.

3.1.3 Phase 3: Mapping of Frequently Used Adjectives on Goldman's Evaluative Aesthetic Terms

A final list of 500 adjectives was prepared representing the most popular adjectives used for describing motorbikes in India. These adjectives were further mapped on Goldman's (1995) eight evaluative aesthetic terms: evaluative, formal, emotional, evocative, behavioral, representational, perceptual, and historical terms (discussed in Chapter 2, Section 2.4.4). Table 3.3 shows the tagging of adjectives on these aesthetic terms. These adjectives were further studied in detail to identify a set of opposite adjectives that could form a semantic differential scale.

TABLE 3.1
Word-Count Study of Seven Motorbike Websites Using Inspyder Insite 2.0 Software

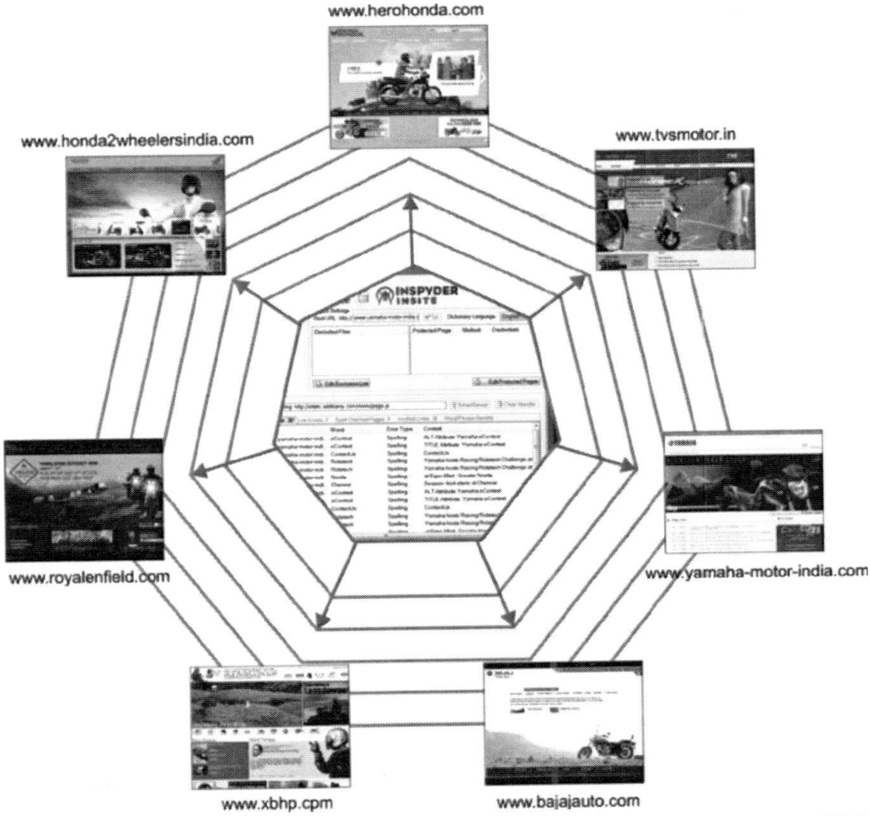

Company Website	Hero Honda	Yamaha	Bajaj	Honda	TVS	Royal Enfield	XBHP	Total
Total no. of words	64,402	9,556	28,452	22,937	99,659	477,756	2,458,330	3,161,092
Unique words	6,703	2,437	5,573	3,379	4,628	25,960	41,530	90,210
Web pages	80	78	124	127	156	905	[a]	>1,470
Links	255	322	385	712	355	1,826	[a]	>3,855
Total adjectives	9,490	1,622	5,024	3,609	7,442	89,633	321,762	438,582
Unique adjectives	1,059	362	848	579	712	2,881	2,549	3,643

[a] Data not recorded.

TABLE 3.2

List of 1,000 Most Frequently Used Adjectives on Motorbike-Related Websites

No.	Adjectives	%	No.	Adjectives	%	No.	Adjectives	%	No.	Adjectives	%
1	motor	4.24	51	excellent	0.35	101	powered	0.21	151	environmental	0.14
2	home	3.74	52	latest	0.35	102	based	0.21	152	several	0.14
3	about	2.17	53	current	0.34	103	kinetic	0.21	153	possible	0.14
4	other	1.54	54	level	0.34	104	impressive	0.20	154	Past	0.14
5	copyright	1.23	55	going	0.33	105	driving	0.20	155	hydraulic	0.14
6	back	1.18	56	down	0.33	106	spare	0.20	156	average	0.13
7	reserved	1.13	57	available	0.33	107	grand	0.20	157	entire	0.13
8	total	1.09	58	select	0.32	108	social	0.20	158	prestigious	0.13
9	after	1.09	59	further	0.32	109	small	0.19	159	managed	0.13
10	private	1.01	60	sure	0.32	110	following	0.19	160	outstanding	0.13
11	even	0.98	61	strong	0.32	111	forward	0.19	161	material	0.13
12	global	0.97	62	state	0.32	112	mono	0.19	162	heavy	0.13
13	only	0.97	63	comfortable	0.32	113	complete	0.18	163	whole	0.12
14	period	0.95	64	full	0.31	114	south	0.18	164	early	0.12
15	like	0.93	65	commercial	0.31	115	smooth	0.18	165	kind	0.12
16	public	0.88	66	single	0.30	116	industrial	0.18	166	rural	0.12
17	good	0.84	67	overall	0.30	117	green	0.18	167	exciting	0.12
18	same	0.83	68	general	0.30	118	senior	0.18	168	manual	0.12
19	royal	0.81	69	continued	0.30	119	technical	0.17	169	related	0.12
20	best	0.80	70	international	0.28	120	called	0.17	170	significant	0.12
21	just	0.80	71	still	0.28	121	everyday	0.17	171	expected	0.12
22	next	0.80	72	advanced	0.28	122	premium	0.17	172	basic	0.12
23	front	0.77	73	light	0.28	123	universal	0.17	173	double	0.12
24	offline	0.76	74	young	0.27	124	safe	0.17	174	East	0.12
25	well	0.74	75	genuine	0.27	125	maximum	0.17	175	clear	0.12
26	then	0.70	76	decided	0.27	126	given	0.17	176	local	0.12
27	through	0.70	77	coming	0.27	127	received	0.17	177	Blue	0.12
28	quality	0.69	78	left	0.27	128	mobile	0.17	178	wanted	0.12
29	very	0.68	79	family	0.27	129	superior	0.16	179	proper	0.12
30	standard	0.62	80	done	0.26	130	metallic	0.16	180	stylish	0.11
31	rear	0.61	81	round	0.25	131	foreign	0.16	181	silver	0.11
32	long	0.59	82	large	0.25	132	super	0.16	182	huge	0.11
33	high	0.55	83	easy	0.25	133	little	0.16	183	West	0.11
34	limited	0.55	84	economic	0.25	134	different	0.16	184	late	0.11
35	under	0.54	85	higher	0.25	135	together	0.16	185	existing	0.11
36	developed	0.53	86	executive	0.25	136	short	0.16	186	printable	0.11
37	stock	0.51	87	corresponding	0.24	137	online	0.15	187	independent	0.11
38	better	0.50	88	special	0.24	138	wide	0.15	188	inside	0.11
39	above	0.50	89	fiscal	0.24	139	paid	0.15	189	running	0.11
40	great	0.48	90	real	0.24	140	various	0.15	190	competitive	0.11
41	national	0.48	91	price	0.23	141	extended	0.15	191	superb	0.11

TABLE 3.2 (*Continued*)
List of 1,000 Most Frequently Used Adjectives on Motorbike-Related Websites

No.	Adjectives	%	No.	Adjectives	%	No.	Adjectives	%	No.	Adjectives	%
42	class	0.47	92	geared	0.23	142	later	0.15	192	individual	0.11
43	free	0.46	93	cool	0.23	143	near	0.15	193	working	0.11
44	financial	0.44	94	sunny	0.22	144	hard	0.15	194	applicable	0.11
45	additional	0.42	95	unique	0.22	145	major	0.15	195	successful	0.11
46	annual	0.42	96	trying	0.22	146	able	0.15	196	ready	0.11
47	corporate	0.40	97	black	0.22	147	able	0.15	197	original	0.10
48	united	0.38	98	important	0.21	148	winning	0.15	198	happy	0.10
49	august	0.37	99	necessary	0.21	149	sporty	0.14	199	covered	0.10
50	previous	0.36	100	open	0.21	150	nice	0.14	200	flat	0.10

No.	Adjectives	%	No.	Adjectives	%	No.	Adjectives	%	No.	Adjectives	%
201	electrical	0.10	251	perfect	0.08	301	actual	0.06	351	continuous	0.05
202	similar	0.10	252	sticky	0.08	302	pickup	0.06	352	relevant	0.05
203	growing	0.10	253	mega	0.08	303	welcome	0.06	353	adequate	0.05
204	fast	0.10	254	minimum	0.08	304	interim	0.06	354	undisputed	0.05
205	north	0.10	255	tired	0.07	305	understanding	0.06	355	neutral	0.05
206	beautiful	0.10	256	proud	0.07	306	packed	0.06	356	mechanical	0.05
207	normal	0.10	257	positive	0.07	307	flexible	0.06	357	fluid	0.05
208	capital	0.10	258	motoring	0.07	308	born	0.06	358	particular	0.05
209	daily	0.10	259	quick	0.07	309	convenient	0.06	359	fell	0.05
210	established	0.09	260	middle	0.07	310	educational	0.06	360	macho	0.05
211	consecutive	0.09	261	central	0.07	311	spent	0.06	361	assistant	0.05
212	sleek	0.09	262	declared	0.07	312	modern	0.06	362	fine	0.05
213	domestic	0.09	263	likely	0.07	313	operational	0.05	363	prior	0.04
214	final	0.09	264	regular	0.07	314	checked	0.05	364	serious	0.04
215	cumulative	0.09	265	friendly	0.07	315	stable	0.05	365	dual	0.04
216	closed	0.09	266	official	0.07	316	finished	0.05	366	fellow	0.04
217	efficient	0.09	267	awesome	0.07	317	extensive	0.05	367	professional	0.04
218	problem	0.09	268	poor	0.07	318	remarkable	0.05	368	intelligent	0.04
219	initial	0.09	269	alone	0.07	319	robust	0.05	369	bottom	0.04
220	stay	0.09	270	subject	0.07	320	proprietary	0.05	370	incredible	0.04
221	headed	0.09	271	slow	0.07	321	integrated	0.05	371	stunning	0.04
222	subsidiary	0.09	272	outside	0.07	322	aware	0.05	372	extreme	0.04
223	recent	0.09	273	steady	0.07	323	telescopic	0.05	373	organized	0.04
224	twin	0.09	274	committed	0.07	324	mounted	0.05	374	narrow	0.04
225	responsible	0.09	275	developing	0.07	325	physical	0.05	375	standing	0.04
226	chief	0.09	276	pretty	0.07	326	dynamic	0.05	376	focused	0.04
227	aggressive	0.09	277	innovative	0.07	327	popular	0.05	377	reliable	0.04
228	extra	0.09	278	regional	0.06	328	contemporary	0.05	378	primary	0.04
229	straight	0.09	279	destination	0.06	329	western	0.05	379	direct	0.04

Continued

TABLE 3.2 (*Continued*)

List of 1,000 Most Frequently Used Adjectives on Motorbike-Related Websites

No.	Adjectives	%	No.	Adjectives	%	No.	Adjectives	%	No.	Adjectives	%
230	political	0.09	280	respective	0.06	330	interesting	0.05	380	potential	0.04
231	human	0.09	281	exclusive	0.06	331	prime	0.05	381	Asian	0.04
232	gone	0.08	282	difficult	0.06	332	sound	0.05	382	essential	0.04
233	worldwide	0.08	283	monthly	0.06	333	legal	0.05	383	opening	0.04
234	constant	0.08	284	core	0.06	334	correct	0.05	384	beat	0.04
235	attractive	0.08	285	active	0.06	335	former	0.05	385	separate	0.04
236	powerful	0.08	286	conventional	0.06	336	rich	0.05	386	stringent	0.04
237	urban	0.08	287	tried	0.06	337	involved	0.05	387	rough	0.04
238	choice	0.08	288	express	0.06	338	becoming	0.05	388	healthy	0.04
239	immediately	0.08	289	prize	0.06	339	fresh	0.05	389	sudden	0.04
240	roll	0.08	290	thinking	0.06	340	dedicated	0.05	390	variable	0.04
241	clean	0.08	291	known	0.06	341	dark	0.05	391	reasonable	0.04
242	cutting	0.08	292	spot	0.06	342	fixed	0.05	392	consequential	0.04
243	amazing	0.08	293	confident	0.06	343	vast	0.05	393	usual	0.04
244	lost	0.08	294	driven	0.06	344	living	0.05	394	bright	0.04
245	automotive	0.08	295	digital	0.06	345	compact	0.05	395	critical	0.04
246	natural	0.08	296	American	0.06	346	idle	0.05	396	flying	0.04
247	moving	0.08	297	consistent	0.06	347	simple	0.05	397	crazy	0.04
248	mean	0.08	298	electric	0.06	348	famous	0.05	398	plastic	0.04
249	cross	0.08	299	changed	0.06	349	worth	0.05	399	brief	0.04
250	flash	0.08	300	multiple	0.06	350	concerned	0.05	400	distinctive	0.04
No.	**Adjectives**	**%**	**No.**	**Adjectives**	**%**	**No.**	**Adjectives**	**%**	**No.**	**Adjectives**	**%**
401	instruction	0.04	451	sufficient	0.03	501	won't	0.03	551	weak	0.02
402	effective	0.04	452	testing	0.03	502	controlled	0.03	552	directional	0.02
403	remaining	0.04	453	valid	0.03	503	routine	0.03	553	memorable	0.02
404	internal	0.04	454	kindly	0.03	504	detailed	0.03	554	soft	0.02
405	wrong	0.04	455	solo	0.03	505	game	0.03	555	damn	0.02
406	wild	0.04	456	wonderful	0.03	506	worst	0.03	556	rare	0.02
407	reverse	0.04	457	external	0.03	507	racy	0.03	557	tubular	0.02
408	certain	0.04	458	phenomenal	0.03	508	prepared	0.03	558	bumpy	0.02
409	medium	0.04	459	sometime	0.03	509	filling	0.03	559	missing	0.02
410	fundamental	0.04	460	historic	0.03	510	expert	0.03	560	evident	0.02
411	common	0.04	461	ample	0.03	511	meritorious	0.03	561	vibrant	0.02
412	ideal	0.04	462	inner	0.03	512	eligible	0.03	562	sensible	0.02
413	himself	0.04	463	affected	0.03	513	steep	0.03	563	hungry	0.02
414	white	0.04	464	passing	0.03	514	ongoing	0.03	564	shut	0.02
415	incorporated	0.04	465	renowned	0.03	515	quarterly	0.03	565	promising	0.02
416	appropriate	0.04	466	splendid	0.03	516	marked	0.03	566	enthusiastic	0.02
417	decent	0.04	467	written	0.03	517	sorry	0.03	567	lean	0.02
418	located	0.04	468	lucky	0.03	518	busy	0.03	568	traveling	0.02

TABLE 3.2 (*Continued*)

List of 1,000 Most Frequently Used Adjectives on Motorbike-Related Websites

No.	Adjectives	%	No.	Adjectives	%	No.	Adjectives	%	No.	Adjectives	%
419	victorious	0.04	469	grown	0.03	519	periodic	0.03	569	lined	0.02
420	warm	0.04	470	inclusive	0.03	520	challenging	0.03	570	engaged	0.02
421	overseas	0.04	471	revolutionary	0.03	521	adopted	0.03	571	capable	0.02
422	useful	0.04	472	budget	0.03	522	touching	0.03	572	stuck	0.02
423	camp	0.04	473	interested	0.03	523	nationwide	0.03	573	prominent	0.02
424	dead	0.04	474	nearby	0.03	524	firm	0.03	574	discerning	0.02
425	masculine	0.04	475	English	0.03	525	ambitious	0.03	575	utility	0.02
426	proven	0.04	476	inspired	0.03	526	placed	0.03	576	mass	0.02
427	unmatched	0.04	477	mere	0.03	527	advance	0.03	577	telling	0.02
428	virtual	0.04	478	associated	0.03	528	sustainable	0.03	578	ranking	0.02
429	forced	0.04	479	pioneering	0.03	529	productive	0.03	579	sophisticated	0.02
430	broken	0.04	480	improper	0.03	530	helical	0.03	580	French	0.02
431	immediate	0.04	481	numerous	0.03	531	winding	0.03	581	strategic	0.02
432	medical	0.04	482	master	0.03	532	administrative	0.03	582	eastern	0.02
433	applied	0.04	483	unprecedented	0.03	533	secondary	0.03	583	minimal	0.02
434	chance	0.04	484	sharp	0.03	534	fair	0.03	584	visible	0.02
435	confused	0.04	485	golden	0.03	535	lovely	0.03	585	excess	0.02
436	loaded	0.04	486	loose	0.03	536	designer	0.03	586	sarcastic	0.02
437	premier	0.03	487	supposed	0.03	537	backward	0.03	587	emotional	0.02
438	favorite	0.03	488	mandatory	0.03	538	minor	0.03	588	privileged	0.02
439	miscellaneous	0.03	489	spacious	0.03	539	accurate	0.03	589	sheer	0.02
440	walking	0.03	490	rapid	0.03	540	rose	0.02	590	reputed	0.02
441	tremendous	0.03	491	experienced	0.03	541	marginal	0.02	591	metro	0.02
442	practical	0.03	492	classic	0.03	542	happening	0.02	592	inspiring	0.02
443	liable	0.03	493	shot	0.03	543	fitting	0.02	593	upcoming	0.02
444	broad	0.03	494	duplicate	0.03	544	substantial	0.02	594	appealing	0.02
445	satisfied	0.03	495	exceptional	0.03	545	aerial	0.02	595	waste	0.02
446	blazing	0.03	496	overwhelming	0.03	546	objective	0.02	596	shaped	0.02
447	fitted	0.03	497	span	0.03	547	impossible	0.02	597	musical	0.02
448	unlikely	0.03	498	automatic	0.03	548	upward	0.02	598	intense	0.02
449	gray	0.03	499	military	0.03	549	valuable	0.02	599	festive	0.02
450	proof	0.03	500	ordinary	0.03	550	sizeable	0.02	600	dusty	0.02

No.	Adjectives	%	No.	Adjectives	%	No.	Adjectives	%	No.	Adjectives	%
601	patent	0.02	651	unable	0.02	701	virgin	0.02	751	pushed	0.01
602	empty	0.02	652	willing	0.02	702	momentous	0.02	752	dangerous	0.01
603	unlike	0.02	653	instant	0.02	703	meaning	0.02	753	collected	0.01
604	enabled	0.02	654	scenic	0.02	704	yearly	0.02	754	closing	0.01
605	distinct	0.02	655	zonal	0.02	705	freezing	0.02	755	quiet	0.01
606	assured	0.02	656	inevitable	0.02	706	formal	0.02	756	formidable	0.01

Continued

TABLE 3.2 (*Continued*)

List of 1,000 Most Frequently Used Adjectives on Motorbike-Related Websites

No.	Adjectives	%	No.	Adjectives	%	No.	Adjectives	%	No.	Adjectives	%
607	visiting	0.02	657	informed	0.02	707	vital	0.02	757	male	0.01
608	fantastic	0.02	658	bound	0.02	708	severe	0.02	758	peak	0.01
609	British	0.02	659	silent	0.02	709	enlightened	0.02	759	regulation	0.01
610	excited	0.02	660	opposite	0.02	710	northern	0.01	760	atmospheric	0.01
611	thumping	0.02	661	outboard	0.02	711	portable	0.01	761	occupied	0.01
612	unread	0.02	662	incomparable	0.02	712	immense	0.01	762	adjacent	0.01
613	surprised	0.02	663	absolute	0.02	713	worn	0.01	763	awake	0.01
614	marine	0.02	664	terrific	0.02	714	greenfield	0.01	764	sweet	0.01
615	dependent	0.02	665	strict	0.02	715	intensive	0.01	765	considerable	0.01
616	yellow	0.02	666	affirmative	0.02	716	lesser	0.01	766	divided	0.01
617	searching	0.02	667	gross	0.02	717	rotary	0.01	767	flagged	0.01
618	rolling	0.02	668	smart	0.02	718	cheap	0.01	768	statutory	0.01
619	chicken	0.02	669	distant	0.02	719	manifold	0.01	769	wanting	0.01
620	refined	0.02	670	comprehensive	0.02	720	merchant	0.01	770	consequent	0.01
621	pressed	0.02	671	outer	0.02	721	sprung	0.01	771	thoughtful	0.01
622	distinguished	0.02	672	triple	0.02	722	accountable	0.01	772	liquid	0.01
623	Sensitive	0.02	673	broke	0.02	723	thrilling	0.01	773	indirect	0.01
624	adjustable	0.02	674	sturdy	0.02	724	conducive	0.01	774	touched	0.01
625	viable	0.02	675	secure	0.02	725	roaring	0.01	775	burning	0.01
626	frequent	0.02	676	plain	0.02	726	breathtaking	0.01	776	boring	0.01
627	weekly	0.02	677	glad	0.02	727	acceptable	0.01	777	Thick	0.01
628	favorable	0.02	678	incomplete	0.02	728	traditional	0.01	778	illustrious	0.01
629	theme	0.02	679	precise	0.02	729	catching	0.01	779	studied	0.01
630	latter	0.02	680	promotional	0.02	730	alternative	0.01	780	unbelievable	0.01
631	poised	0.02	681	ripping	0.02	731	afraid	0.01	781	fleet	0.01
632	signal	0.02	682	remote	0.02	732	spirited	0.01	782	mutual	0.01
633	collect	0.02	683	beige	0.02	733	articulate	0.01	783	worthy	0.01
634	indigenous	0.02	684	grayish	0.02	734	scared	0.01	784	exact	0.01
635	disabled	0.02	685	obvious	0.02	735	ultimate	0.01	785	gold	0.01
636	vocational	0.02	686	memorial	0.02	736	prospective	0.01	786	tangible	0.01
637	nuts	0.02	687	congested	0.02	737	familiar	0.01	787	worse	0.01
638	rank	0.02	688	honorable	0.02	738	careful	0.01	788	astounding	0.01
639	maiden	0.02	689	eminent	0.02	739	typical	0.01	789	rusty	0.01
640	offset	0.02	690	animal	0.02	740	hurt	0.01	790	foremost	0.01
641	upper	0.02	691	celebrated	0.02	741	adult	0.01	791	frugal	0.01
642	extraordinary	0.02	692	custom	0.02	742	famed	0.01	792	successive	0.01
643	relentless	0.02	693	flush	0.02	743	massive	0.01	793	resistant	0.01
644	sunset	0.02	694	spiral	0.02	744	comparable	0.01	794	passive	0.01
645	ordered	0.02	695	furnished	0.02	745	durable	0.01	795	dominant	0.01
646	crowded	0.02	696	visual	0.02	746	intrinsic	0.01	796	enjoyable	0.01
647	crucial	0.02	697	pure	0.02	747	lightning	0.01	797	lazy	0.01

TABLE 3.2 (*Continued*)

List of 1,000 Most Frequently Used Adjectives on Motorbike-Related Websites

No.	Adjectives	%	No.	Adjectives	%	No.	Adjectives	%	No.	Adjectives	%
648	pleasant	0.02	698	female	0.02	748	civil	0.01	798	unclaimed	0.01
649	optimistic	0.02	699	German	0.02	749	intermediate	0.01	799	legendary	0.01
650	relaxed	0.02	700	mighty	0.02	750	honorary	0.01	800	draft	0.01

No.	Adjectives	%	No.	Adjectives	%	No.	Adjectives	%	No.	Adjectives	%
801	compromising	0.01	851	defective	0.01	901	slight	0.01	951	negligent	0.01
802	learned	0.01	852	pilot	0.01	902	exacting	0.01	952	purplish	0.01
803	diverse	0.01	853	maroon	0.01	903	square	0.01	953	shaky	0.01
804	youthful	0.01	854	uncomfortable	0.01	904	biting	0.01	954	unparalleled	0.01
805	raring	0.01	855	accidental	0.01	905	suited	0.01	955	alert	0.01
806	underway	0.01	856	rival	0.01	906	worried	0.01	956	desirable	0.01
807	southern	0.01	857	governing	0.01	907	strange	0.01	957	fixated	0.01
808	glorious	0.01	858	brilliant	0.01	908	animated	0.01	958	interior	0.01
809	radiant	0.01	859	varied	0.01	909	sporting	0.01	959	inviting	0.01
810	chill	0.01	860	supporting	0.01	910	delighted	0.01	960	judicious	0.01
811	intellectual	0.01	861	absorbing	0.01	911	everlasting	0.01	961	tall	0.01
812	tiny	0.01	862	fallen	0.01	912	acting	0.01	962	modest	0.01
813	exemplary	0.01	863	retaining	0.01	913	chilly	0.01	963	compulsive	0.01
814	dairy	0.01	864	artificial	0.01	914	wearing	0.01	964	sculptured	0.01
815	European	0.01	865	spotted	0.01	915	outdated	0.01	965	steadfast	0.01
816	prosperous	0.01	866	weird	0.01	916	keen	0.01	966	systematic	0.01
817	ingrained	0.01	867	invaluable	0.01	917	surrounding	0.01	967	democratic	0.01
818	thin	0.01	868	giant	0.01	918	stressed	0.01	968	occasional	0.01
819	slack	0.01	869	colored	0.01	919	rewarding	0.01	969	harmful	0.01
820	matching	0.01	870	forthcoming	0.01	920	situated	0.01	970	armed	0.01
821	raised	0.01	871	orange	0.01	921	funny	0.01	971	societal	0.01
822	concentrated	0.01	872	graduated	0.01	922	striking	0.01	972	yielding	0.01
823	unauthorized	0.01	873	exposed	0.01	923	rigorous	0.01	973	thorough	0.01
824	academic	0.01	874	rocky	0.01	924	partial	0.01	974	noted	0.01
825	crash	0.01	875	compulsory	0.01	925	flip	0.01	975	veterinary	0.01
826	renewed	0.01	876	sick	0.01	926	utmost	0.01	976	stupid	0.01
827	pleased	0.01	877	thrilled	0.01	927	knowing	0.01	977	regulatory	0.01
828	inherent	0.01	878	sleepy	0.01	928	loud	0.01	978	cooperative	0.01
829	passionate	0.01	879	inclined	0.01	929	effortless	0.01	979	slippery	0.01
830	historical	0.01	880	African	0.01	930	optional	0.01	980	determined	0.01
831	histrionic	0.01	881	random	0.01	931	countrywide	0.01	981	disappointed	0.01
832	confirmed	0.01	882	rugged	0.01	932	beneficial	0.01	982	unknown	0.01
833	expedient	0.01	883	nimble	0.01	933	eager	0.01	983	governmental	0.01
834	propitious	0.01	884	equitable	0.01	934	uphill	0.01	984	innate	0.01
835	iron	0.01	885	handsome	0.01	935	baby	0.01	985	reformist	0.01

Continued

TABLE 3.2 (*Continued*)

List of 1,000 Most Frequently Used Adjectives on Motorbike-Related Websites

No.	Adjectives	%	No.	Adjectives	%	No.	Adjectives	%	No.	Adjectives	%
836	pending	0.01	886	alive	0.01	936	enduring	0.01	986	unspeakable	0.01
837	optimal	0.01	887	dense	0.01	937	secret	0.01	987	unusual	0.01
838	swift	0.01	888	enormous	0.01	938	bored	0.01	988	opposed	0.01
839	settled	0.01	889	managerial	0.01	939	supreme	0.01	989	dirty	0.01
840	destined	0.01	890	trendy	0.01	940	inexhaustible	0.01	990	magic	0.01
841	agile	0.01	891	tied	0.01	941	unprepared	0.01	991	futuristic	0.01
842	notable	0.01	892	incidental	0.01	942	fabulous	0.01	992	noteworthy	0.01
843	resigned	0.01	893	voluminous	0.01	943	confidential	0.01	993	traveled	0.01
844	principal	0.01	894	Latin	0.01	944	harsh	0.01	994	utilitarian	0.01
845	spectacular	0.01	895	precious	0.01	945	unfamiliar	0.01	995	rushed	0.01
846	restricted	0.01	896	talented	0.01	946	unattended	0.01	996	proverbial	0.01
847	balanced	0.01	897	pleasing	0.01	947	crap	0.01	997	handy	0.01
848	accessible	0.01	898	refreshing	0.01	948	unpaid	0.01	998	intent	0.01
849	dear	0.01	899	owing	0.01	949	unfair	0.01	999	dramatic	0.01
850	educated	0.01	900	downhill	0.01	950	cohesive	0.01	1000	dramatic	0.01

And finally, the following sets of seven opposite adjectives (14 adjectives in all) were identified from this study:

(1) Formal (2) Emotional
(3) Professional (4) Sporty
(5) Extraordinary (6) Popular
(7) Mean (8) Friendly
(9) Aggressive (10) Comfortable
(11) Sharp (12) Heavy
(13) International (14) Traditional

Opposite adjectives were selected from different evaluative aesthetic categories to ensure that each adjective represented a distinct personality variable or trait, related to motorbikes in India. The identified 14 personality-describing adjectives are hereby referred to as "personality variables" (Figure 3.1).

The 14 personality variables were used to develop a self-reporting instrument to measure the biker–bike personality. Figure 3.2 shows the developed instrument, which is divided into two parts. The first part of the instrument can be used for reporting the respondent's (biker's) own personality, on the 14 personality variables, on a five-point rating scale (5—strongly agree, 4—agree, 3—can't say, 2—disagree, 1—strongly disagree). And the second part of the instrument can be used for reporting the bike's personality on each of

TABLE 3.3 (See color insert)

List of 500 Motorbike-Describing Adjectives in India

	Formal		*Evocative*		*Perceptual*		*Representational*		
	Historical		*Emotional*		*Behavoral*		*Broadly Evaluative*		
No.	**Adjectives**	**No.**	**Adjectives**	**No.**	**Adjectives**	**No.**	**Adjectives**	**No.**	**Adjectives**
1	Global	51	Universal	101	Mega	151	Simple	201	Decent
2	Royal	52	Safe	102	Proud	152	Famous	202	Victorious
3	Best	53	Maximum	103	Positive	153	Worth	203	Warm
4	Next	54	Superior	104	Motoring	154	Continuous	204	Useful
5	Front	55	Metallic	105	Quick	155	Undisputed	205	Dead
6	Quality	56	Super	106	Middle	156	Neutral	206	Masculine
7	Very	57	Different	107	Regular	157	Mechanical	207	Unmatched
8	Standard	58	Wide	108	**Friendly**	158	Fluid	208	Forced
9	Rear	59	Hard	109	Awesome	159	Macho	209	Immediate
10	Long	60	Winning	110	Poor	160	Serious	210	Confused
11	High	61	**Sporty**	111	Slow	161	Dual	211	Loaded
12	Better	62	Nice	112	Steady	162	**Professional**	212	Favourite
13	Great	63	Environmental	113	Committed	163	Intelligent	213	Tremendous
14	National	64	Prestigious	114	Pretty	164	Incredible	214	Practical
15	Class	65	Outstanding	115	Innovative	165	Stunning	215	Satisfied
16	Excellent	66	**Heavy**	116	Regional	166	Extreme	216	Blazing
17	Latest	67	Rural	117	Exclusive	167	Narrow	217	Grey
18	Sure	68	Exciting	118	Difficult	168	Standing	218	Sufficient
19	Strong	69	Basic	119	Core	169	Focused	219	Wonderful
20	**Comfortable**	70	Blue	120	Active	170	Reliable	220	Phenomenal
21	Full	71	Stylish	121	Conventional	171	Potential	221	Splendid
22	**International**	72	Silver	122	Express	172	Asian	222	Revolutionary
23	Advanced	73	Huge	123	Thinking	173	Essential	223	Interested
24	Light	74	Running	124	Confident	174	Beat	224	English
25	Young	75	Competitive	125	Digital	175	Separate	225	Inspired
26	Genuine	76	Superb	126	American	176	Stringent	226	Associated
27	Family	77	Successful	127	Consistent	177	Rough	227	Numerous
28	Round	78	Ready	128	Multiple	178	Healthy	228	**Sharp**
29	Large	79	Happy	129	Born	179	Sudden	229	Golden
30	Easy	80	Growing	130	Convenient	180	Variable	230	Spacious
31	Economic	81	Fast	131	Modern	181	Reasonable	231	Rapid
32	Higher	82	Beautiful	132	Stable	182	Consequential	232	Classic
33	Executive	83	Sleek	133	Extensive	183	Usual	233	Exceptional
34	Special	84	Efficient	134	Remarkable	184	Bright	234	Overwhelming
35	Real	85	Chief	135	Robust	185	Critical	235	Automatic
36	Price	86	**Aggressive**	136	Integrated	186	Flying	236	Military
37	Cool	87	Extra	137	Physical	187	Crazy	237	Ordinary
38	Unique	88	Attractive	138	Dynamic	188	Plastic	238	Game
39	Black	89	Powerful	139	**Popular**	189	Brief	239	Worst
40	Important	90	Urban	140	Contemporary	190	Distinctive	240	Racy
41	Necessary	91	Choice	141	Western	191	Effective	241	Expert
42	Powered	92	Immediately	142	Interesting	192	Wrong	242	Eligible
43	Kinetic	93	Clean	143	Prime	193	Wild	243	Periodic
44	Impressive	94	Amazing	144	Sound	194	Certain	244	Touching
45	Driving	95	Automotive	145	Rich	195	Fundamental	245	Nationwide
46	Grand	96	Natural	146	Fresh	196	Common	246	Firm
47	Small	97	Moving	147	Dedicated	197	Ideal	247	Ambitious
48	Smooth	98	**Mean**	148	Dark	198	Himself	248	Advance
49	Green	99	Flash	149	Compact	199	White	249	Sustainable
50	Premium	100	Perfect	150	Idle	200	Appropriate	250	Fair

Continued

TABLE 3.3 (*Continued*) (See color insert)

List of 500 Motorbike-Describing Adjectives in India

	Formal		*Evocative*		*Perceptual*		*Representational*		
	Historical		*Emotional*		*Behavoral*		*Broadly Evaluative*		
No.	**Adjectives**	**No.**	**Adjectives**	**No.**	**Adjectives**	**No.**	**Adjectives**	**No.**	**Adjectives**
251	Lovely	301	British	351	Virgin	401	Youthful	451	Delighted
252	Designer	302	Excited	352	Momentous	402	Raring	452	Everlasting
253	Accurate	303	Thumping	353	Meaning	403	Glorious	453	Outdated
254	Happening	304	Surprised	354	**Formal**	404	Radiant	454	Funny
255	Substantial	305	Yellow	355	Enlightened	405	Chill	455	Striking
256	Impossible	306	Refined	356	Immense	406	European	456	Loud
257	Sizeable	307	Distinguished	357	Intensive	407	Prosperous	457	Effortless
258	Weak	308	Sensitive	358	Cheap	408	Thin	458	Countrywide
259	Directional	309	Adjustable	359	Breathtaking	409	Matching	459	Eager
260	Soft	310	Frequent	360	Acceptable	410	Concentrated	460	Uphill
261	Rare	311	Favourable	361	**Traditional**	411	Unauthorized	461	Baby
262	Evident	312	Theme	362	Afraid	412	Academic	462	Enduring
263	Vibrant	313	Disabled	363	Spirited	413	Pleased	463	Secret
264	Sensible	314	**Extraordinary**	364	Scared	414	Passionate	464	Bored
265	Hungry	315	Relentless	365	Ultimate	415	Confirmed	465	Supreme
266	Promising	316	Pleasant	366	Prospective	416	Swift	466	Inexhaustible
267	Enthusiastic	317	Relaxed	367	Careful	417	Agile	467	Unprepared
268	Lean	318	Instant	368	Typical	418	Principal	468	Fabulous
269	Travelling	319	Inevitable	369	Adult	419	Spectacular	469	Confidential
270	Engaged	320	Opposite	370	Massive	420	Restricted	470	Harsh
271	Prominent	321	Incomparable	371	Comparable	421	Dear	471	Unfamiliar
272	Utility	322	Absolute	372	Durable	422	Pilot	472	Unattended
273	Mass	323	Terrific	373	Intrinsic	423	Maroon	473	Unfair
274	Ranking	324	Affirmative	374	Lightning	424	Uncomfortable	474	Cohesive
275	Sophisticated	325	Smart	375	Honorary	425	Accidental	475	Alert
276	French	326	Comprehensive	376	Dangerous	426	Rival	476	Desirable
277	Strategic	327	Sturdy	377	Quiet	427	Brilliant	477	Interior
278	Eastern	328	Secure	378	Male	428	Absorbing	478	Inviting
279	Minimal	329	Plain	379	Peak	429	Weird	479	Tall
280	Visible	330	Glad	380	Adjacent	430	Coloured	480	Modest
281	Excess	331	Incomplete	381	Sweet	431	Orange	481	Compulsive
282	Sarcastic	332	Precise	382	Consequent	432	Rocky	482	Sculptured
283	**Emotional**	333	Promotional	383	Touched	433	Compulsory	483	Steadfast
284	Privileged	334	Ripping	384	Boring	434	Thrilled	484	Systematic
285	Sheer	335	Remote	385	Thick	435	African	485	Democratic
286	Reputed	336	Beige	386	Illustrious	436	Rugged	486	Occasional
287	Metro	337	Grayish	387	Unbelievable	437	Handsome	487	Harmful
288	Inspiring	338	Obvious	388	Worthy	438	Alive	488	Armed
289	Upcoming	339	Memorial	389	Gold	439	Dense	489	Stupid
290	Appealing	340	Congested	390	Worse	440	Enormous	490	Cooperative
291	Shaped	341	Honourable	391	Astounding	441	Trendy	491	Disappointed
292	Musical	342	Animal	392	Successive	442	Latin	492	Unusual
293	Intense	343	Celebrated	393	Passive	443	Precious	493	Dirty
294	Festive	344	Custom	394	Dominant	444	Pleasing	494	Magic
295	Empty	345	Spiral	395	Enjoyable	445	Refreshing	495	Futuristic
296	Unlike	346	Visual	396	Lazy	446	Slight	496	Noteworthy
297	Enabled	347	Pure	397	Unclaimed	447	Exacting	497	Utilitarian
298	Distinct	348	Female	398	Legendary	448	Strange	498	Rushed
299	Assured	349	German	399	Compromising	449	Animated	499	Handy
300	Fantastic	350	Mighty	400	Diverse	450	Sporting	500	Dramatic

FIGURE 1.1
Motorbike is an expressionf rider's emotions. (From Sunil Gupta of xbhp.com, personal communication, Jan. 27, 2013.)

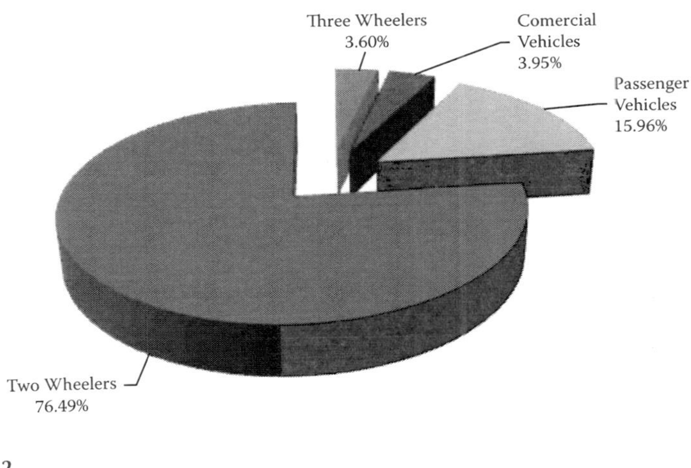

FIGURE 1.2
Category-wise market share of automobiles in India in 2008–2009. (Retrieved May 12, 2009, from http://www.siamindia.com/scripts/market-share.aspx.)

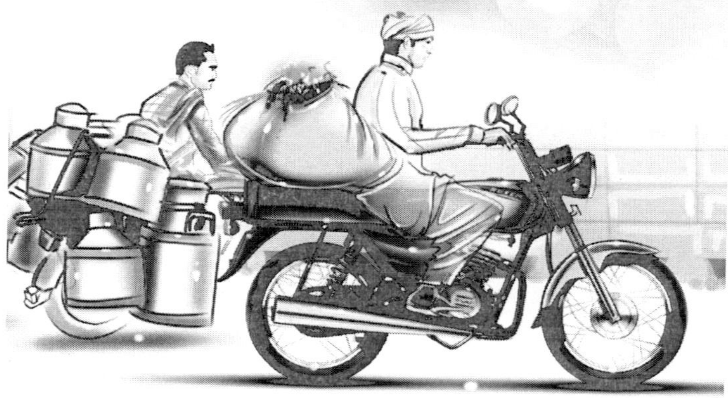

FIGURE 1.4
Pictorial representation of the bread-and-butter biking segment in India. (Illustration by Debidutt Gharai, personal communication, Jan. 28, 2013.)

FIGURE 1.5
Pictorial representation of the travel-and-comfort biking segment in India. (Illustration by Debidutt Gharai, personal communication, Jan. 28, 2013.)

FIGURE 1.6
Pictorial representation of the social-networking biking segment in India. (Illustration by Debidutt Gharai, personal communication, Jan. 28, 2013.)

FIGURE 1.7
Pictorial representation of the dexterous and exploratory biking segment in India. (Illustration by Debidutt Gharai, personal communication, Jan. 30, 2013.)

FIGURE 1.8
Pictorial representation of the high-esteem biking segment in India. (Illustration by Debidutt Gharai, personal communication, Jan. 30, 2013.)

FIGURE 1.9
Pictorial representation of the religious biking segment in India. (Illustration by Debidutt Gharai, personal communication, Jan. 30, 2013.)

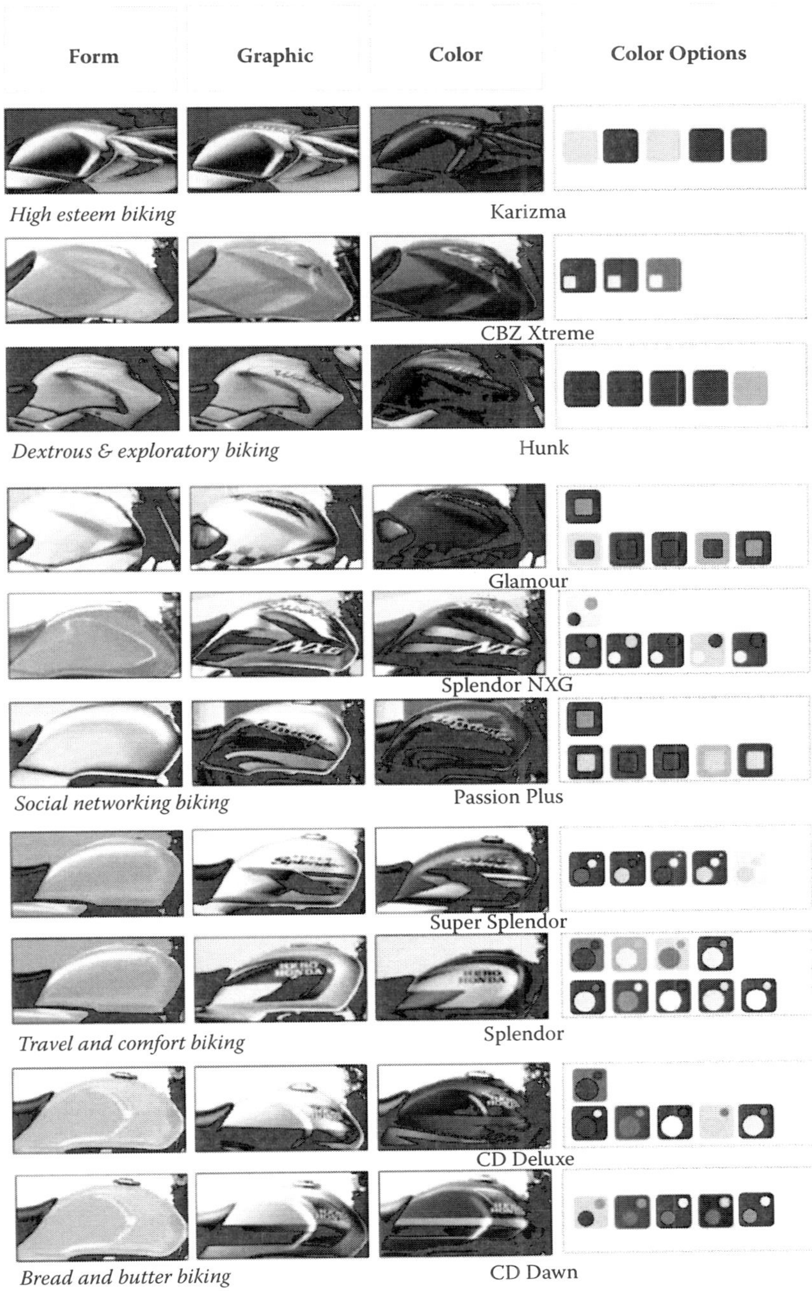

FIGURE 1.10
Visual comparison of aesthetic characteristics (form, graphics, and colors) of motorbikes representing different biking segments.

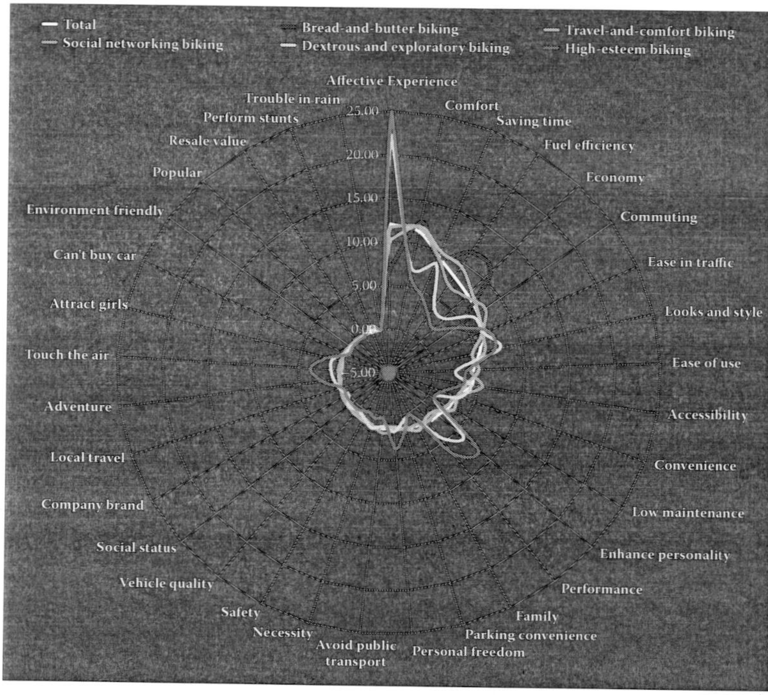

FIGURE 4.2
Comparison between biking needs for each biking segment.

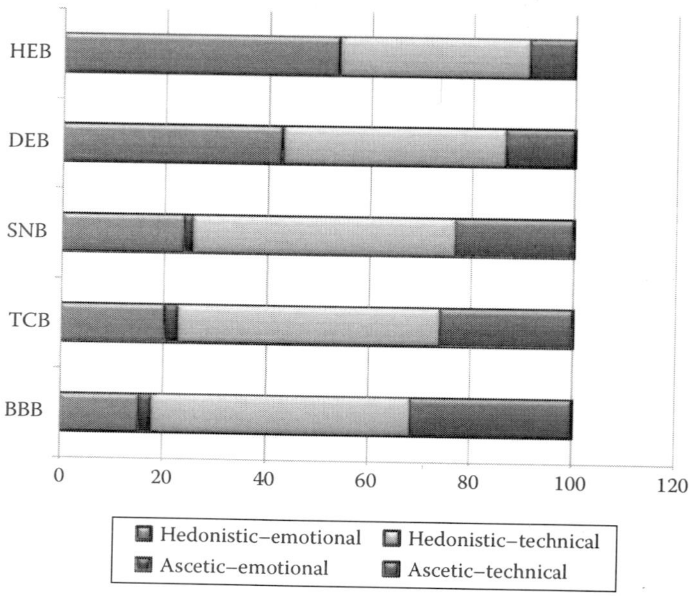

FIGURE 4.4
Percentage of H-E-A-T needs associated with different biking segments.

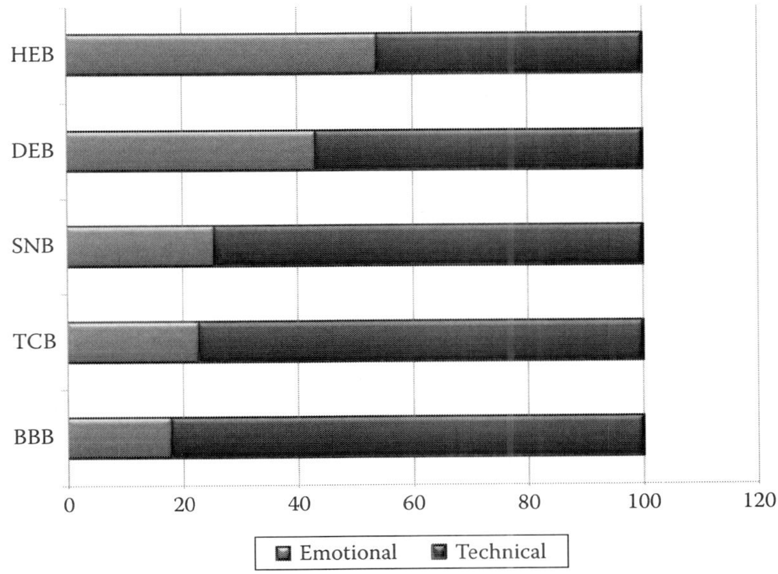

FIGURE 4.5
Comparison of biking needs associated with different biking segments on the emotional–technical dimensions.

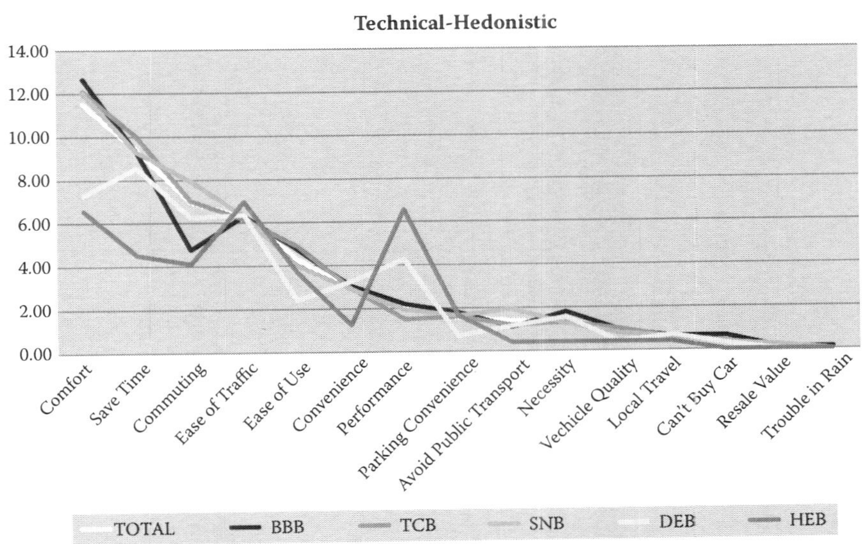

FIGURE 4.6
Technical–hedonistic biking needs associated with different biking segments.

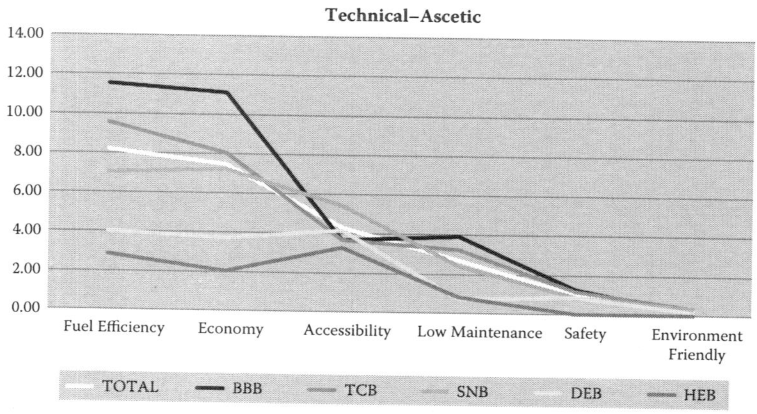

FIGURE 4.7
Technical–ascetic biking needs associated with different biking segments.

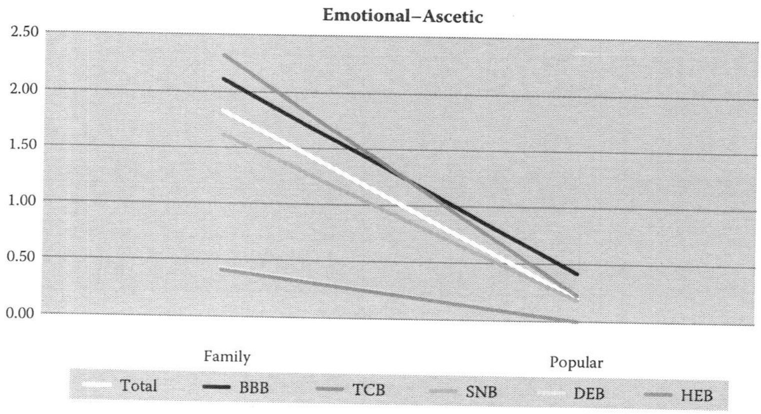

FIGURE 4.8
Emotional–ascetic biking needs associated with different biking segments.

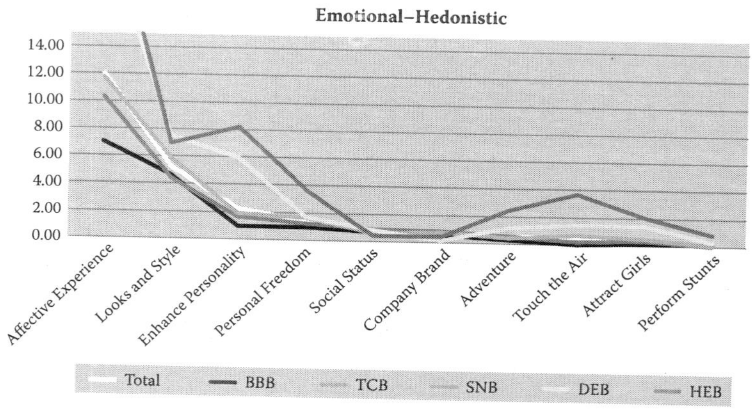

FIGURE 4.9
Emotional–hedonistic biking needs associated with different biking segments.

BBB: Bread and Butter Biking
TCB: Travel and Comfort Biking
SNB: Social Networking Biking
DEB: Dextrous and Exploratory Biking
HEB: High Esteem Biking

	Total	BBB	TCB	SNB	DEB	HEB
Relax	2.13	2.13	0.56	1.25	8.50	0.00
Interest	7.47	21.28	3.07	5.64	17.65	6.67
Enjoy	6.94	38.30	2.23	4.39	10.46	15.00
Love	55.28	29.79	56.42	65.20	46.41	38.33
Passion	27.21	4.26	37.71	23.51	16.99	28.33
Thrill	0.96	4.26	0.00	0.00	0.00	11.67

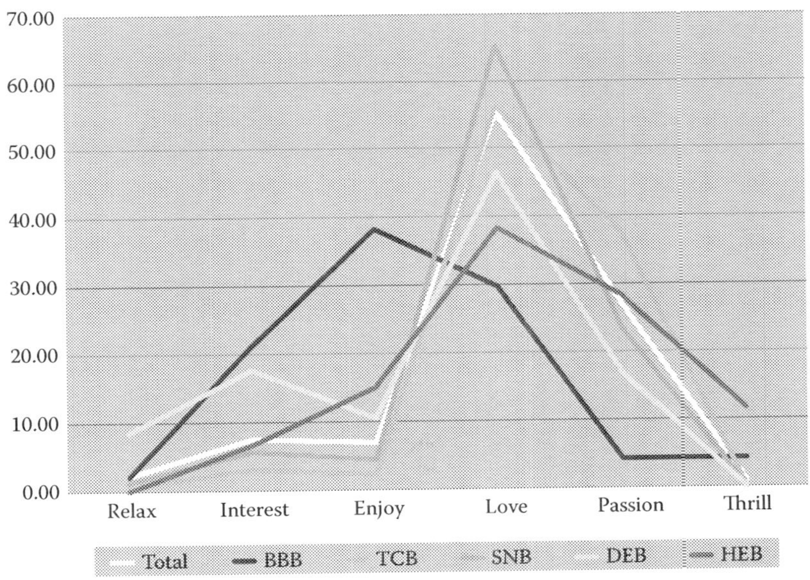

FIGURE 4.11
Affective experience related to each biking segment.

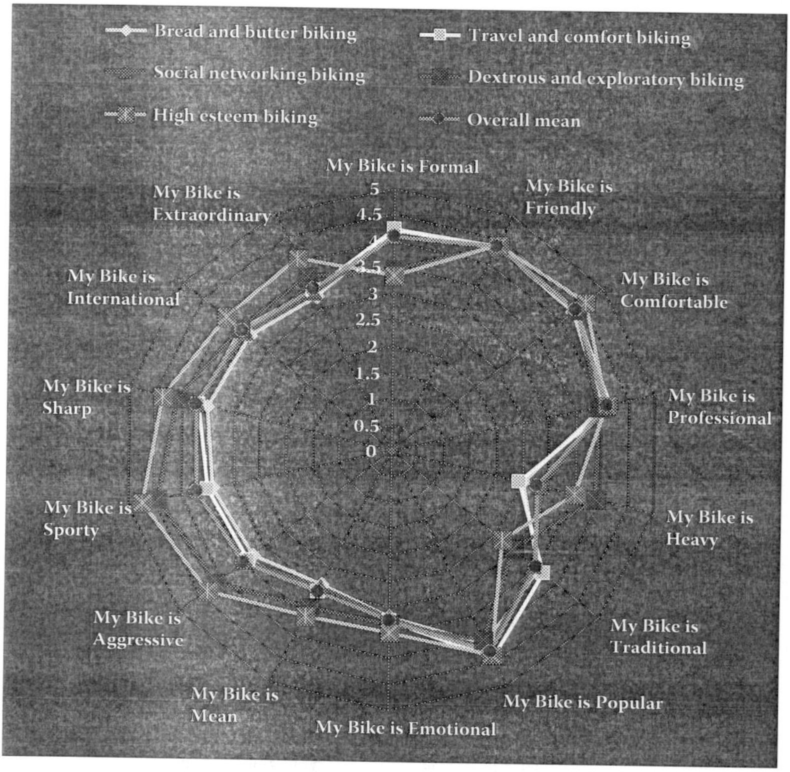

FIGURE 5.1
Mean ratings for bike personalities associated with each biking segment on each personality variable.

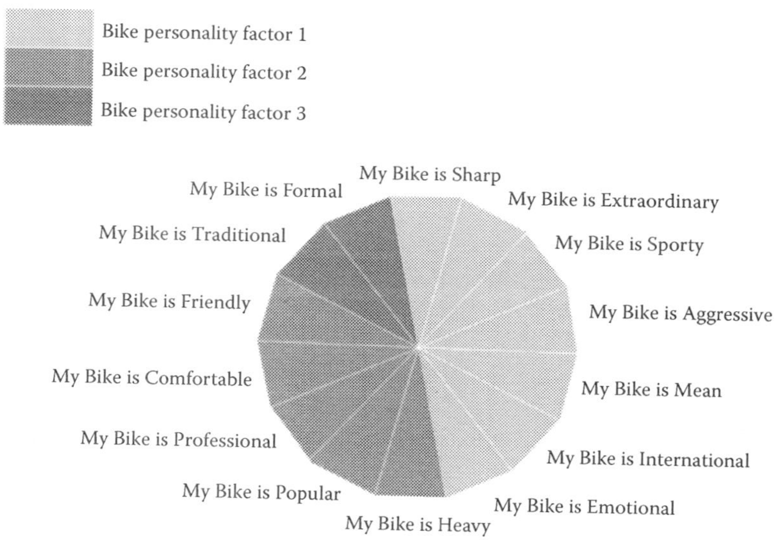

FIGURE 5.2
Bike personality factors.

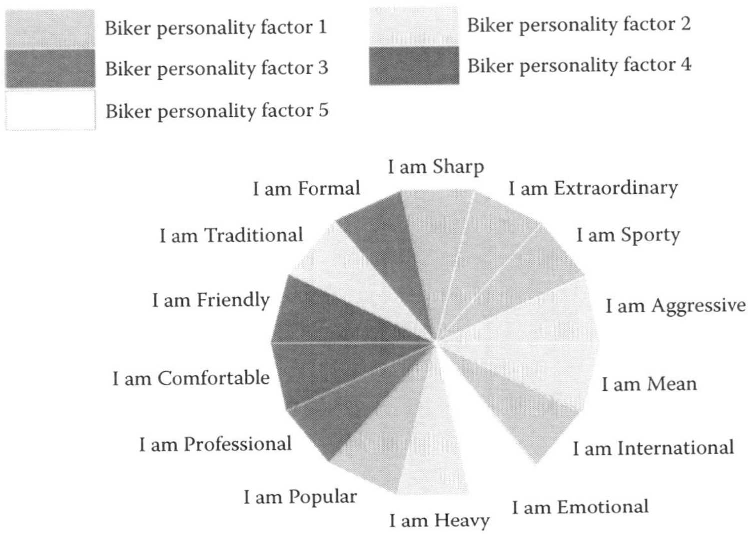

FIGURE 5.3
Biker personality factors.

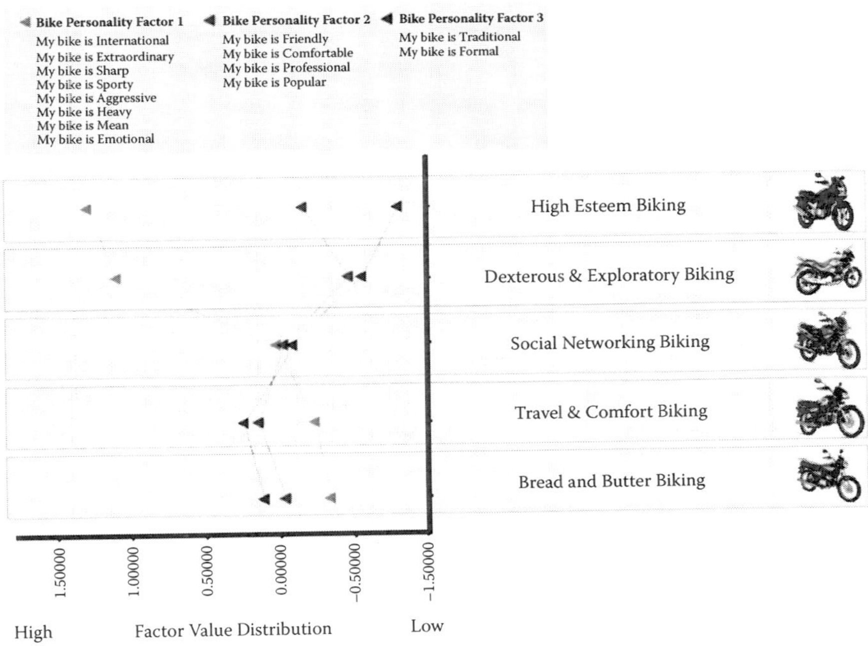

FIGURE 5.4
Multi–scatter plot showing the bike personality associated with different biking segments.

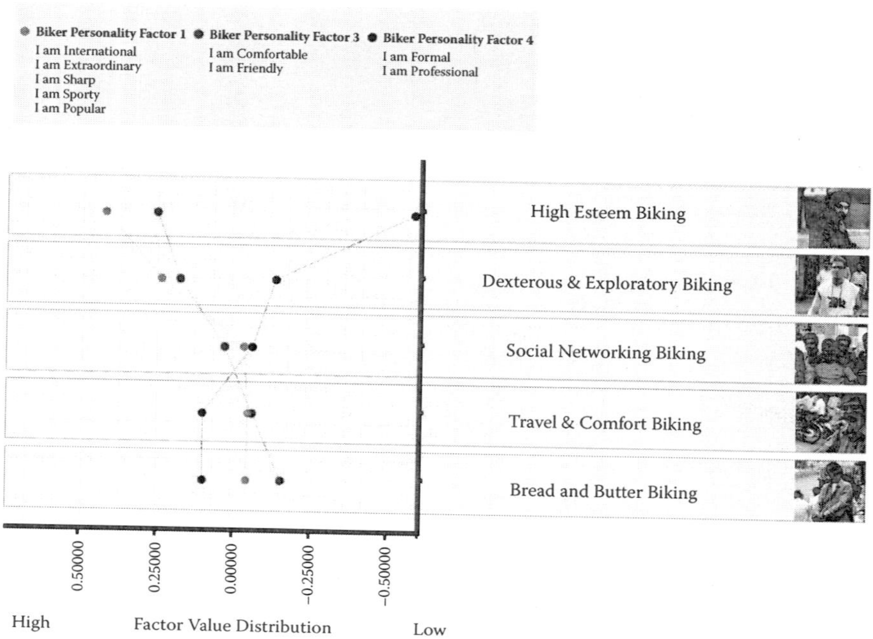

FIGURE 5.5
Multi–scatter plot showing the biker personality associated with different biking segments.

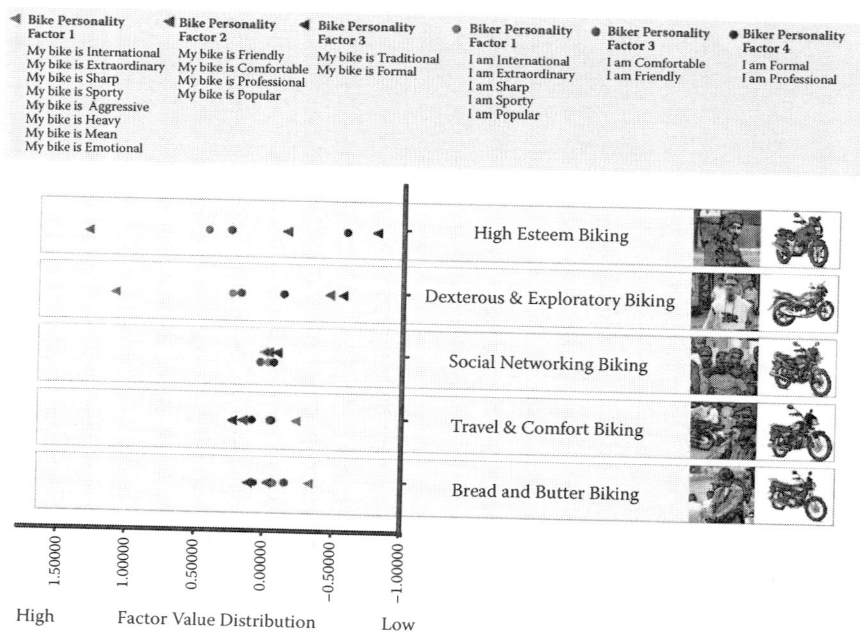

FIGURE 5.6
Multi–scatter plot showing the emotive quality of biking segments in India.

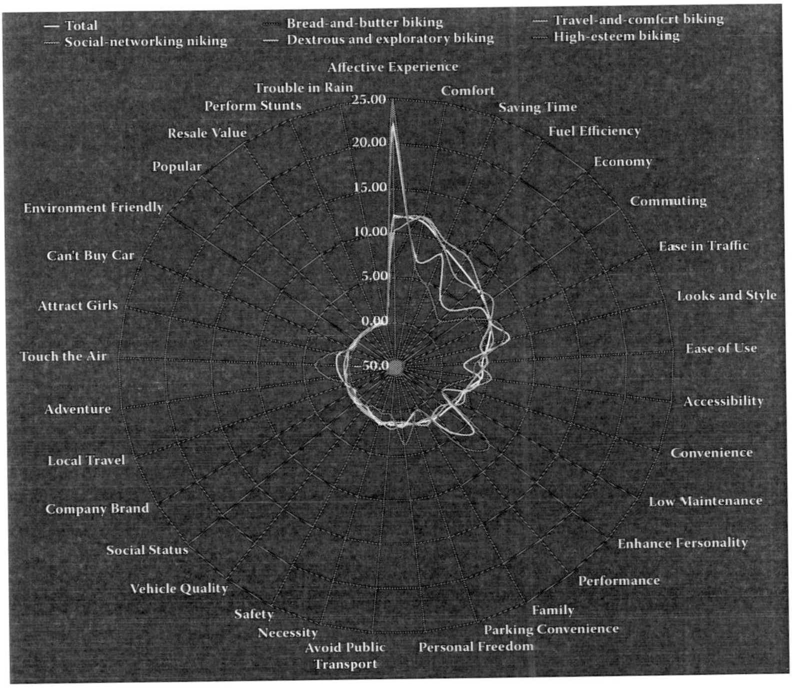

FIGURE 6.1
Comparison between biking needs for each biking segment.

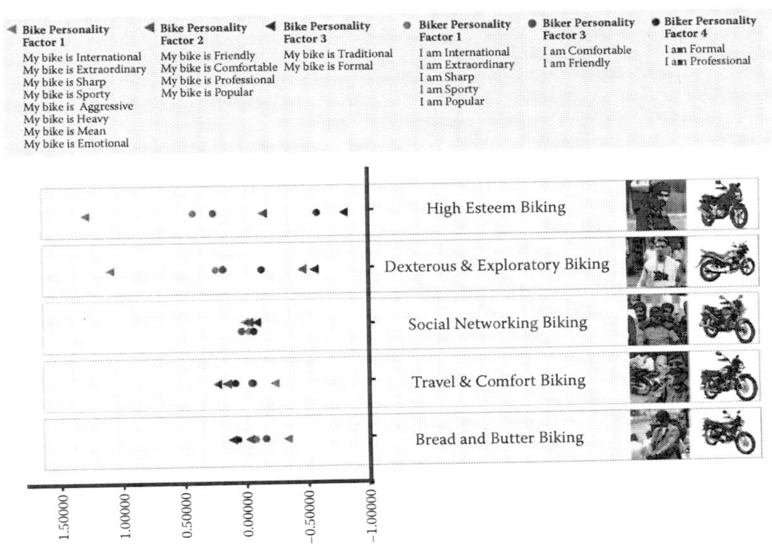

FIGURE 6.2
Multi–scatter plot showing the emotive quality of biking segments in India.

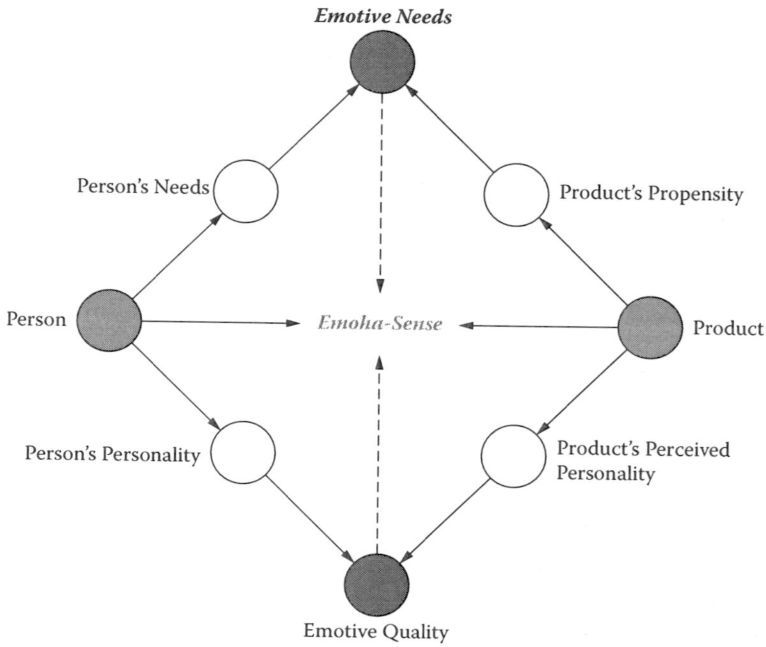

FIGURE 6.4
Model for *emoha*-based product styling.

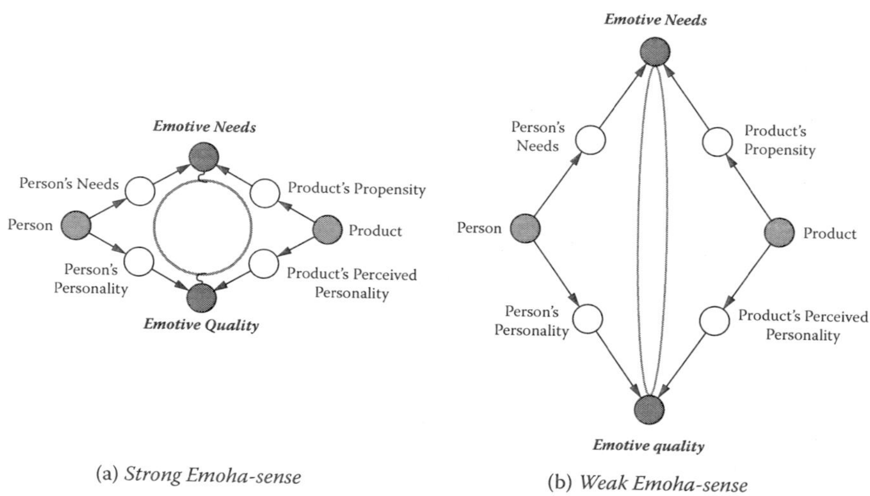

FIGURE 6.5
Depiction of strong and weak *emoha-sense*.

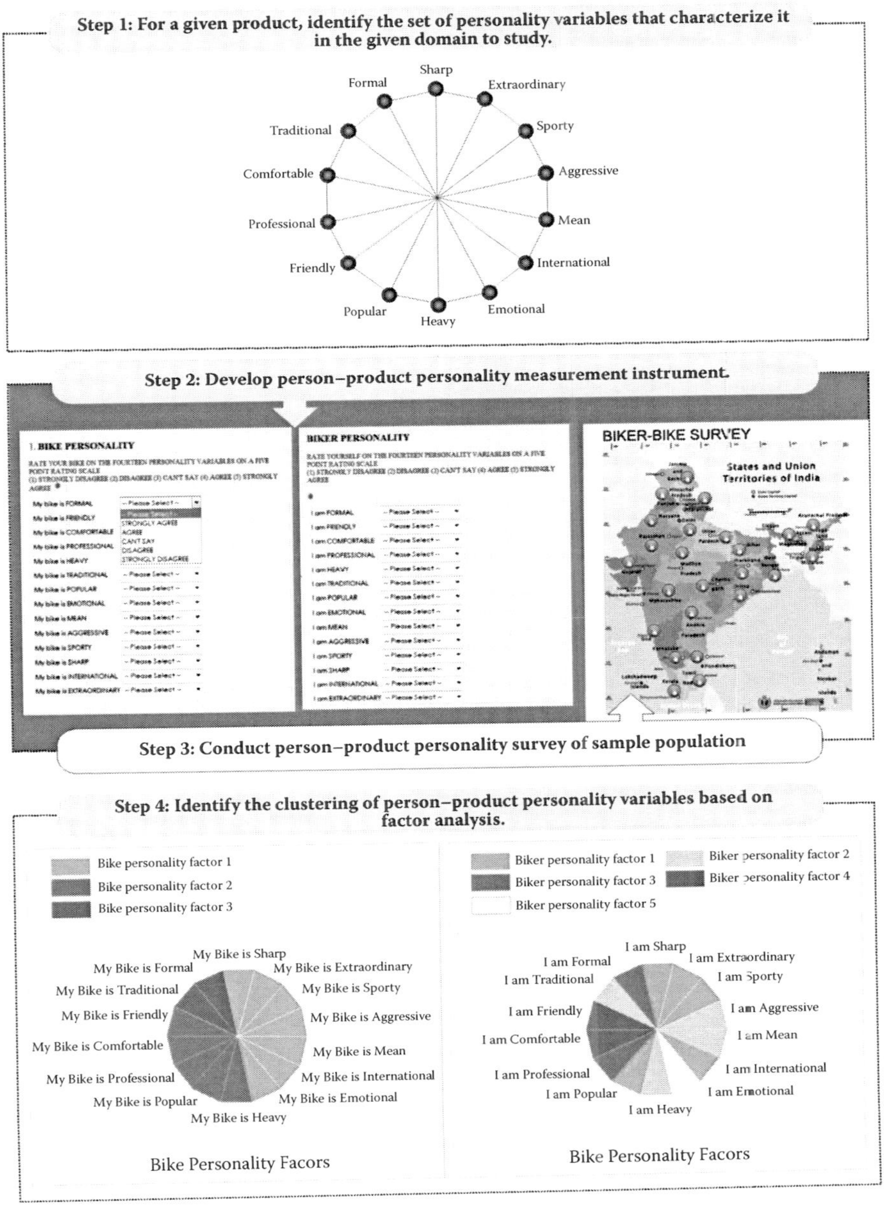

FIGURE 6.6
Application of the "design method for *emoha*-based product styling" in the context of present study of motorbikes in India.

Continued

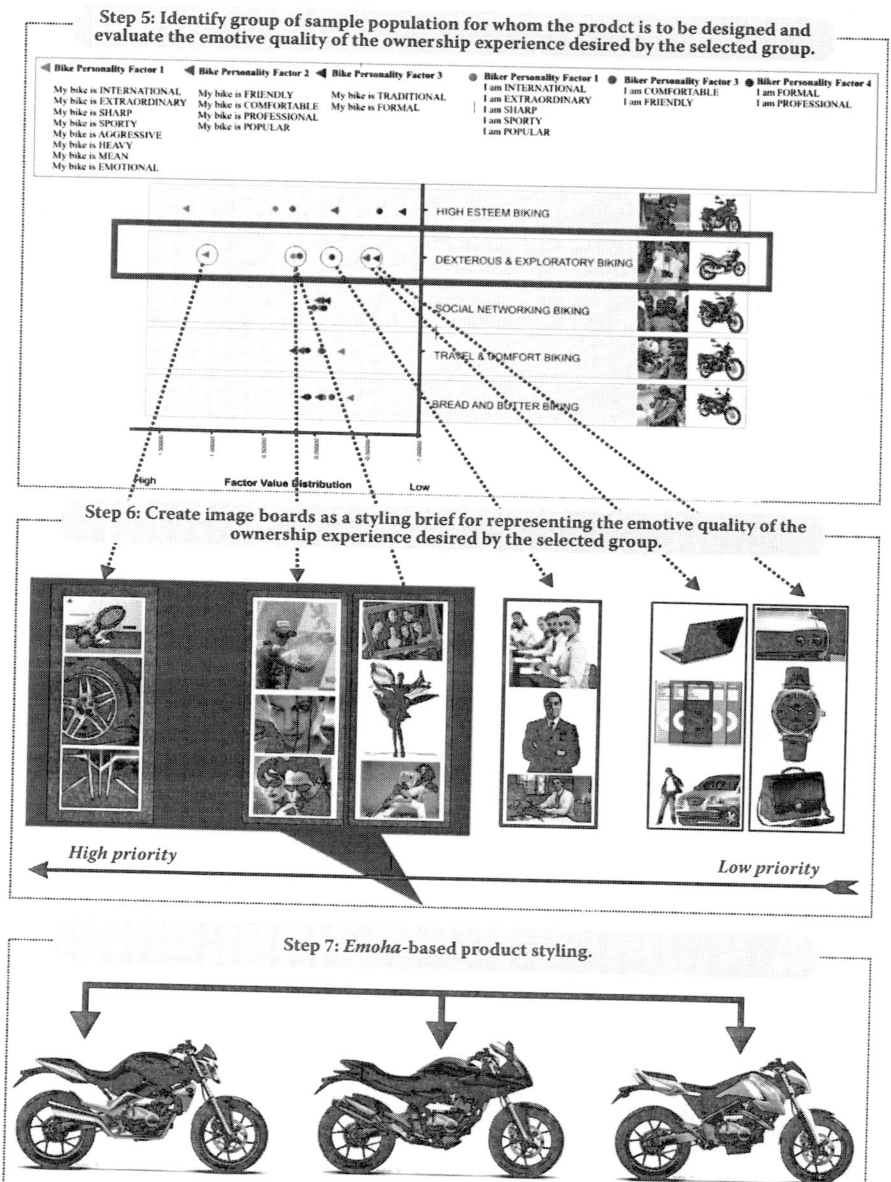

FIGURE 6.6 (*Continued*)
Application of the "design method for *emoha*-based product styling" in the context of present study of motorbikes in India.

TABLE 1.2

Classification of Indian Motorbikes on the Five Biking Segments

Hero Honda	Bajaj	Honda	TVS	Yamaha	Suzuki	Kinetic	Royal Enfield
CD-Dawn	XCD 125	Unicorn	Star	Alba	Heat	Stryker	Bullet Machismo
CD-Deluxe	Platina	Shine	Star City	G5	Zeus 125	Comet	Bullet Machismo 500
Splendor Plus	Discover	CBF Stunner	Star Sport	Crux	Hayabusa	Aquila	Thunderbird
Splendor Nxg	Pulsar DTS-i		Flame	Gladiator			Bullet Electra 5S
Passion Plus	Pulsar 200 DTS-i		Apache RTR 160	MT-01			Bullet Electra
Super Splendor	Pulsar 220 DTS-Fi			YZF-R1			Bullet 350
Glamour	Avenger DTS-i			FZ1			
Glamour Fi	CT 100			YZF R15			
Achiever							
Hunk							
CBZ Xtreme							
Karizma							

High Esteem Biking
Dextrous and Exploratory Biking
Social Networking Biking
Travel and Comfort Biking
Bread and Butter Biking

TABLE 3.3

List of 500 Motorbike-Describing Adjectives in India

	Formal		*Evocative*		*Perceptual*		*Representational*		
	Historical		*Emotional*		*Behavoral*		*Broadly Evaluative*		
No.	**Adjectives**	**No.**	**Adjectives**	**No.**	**Adjectives**	**No.**	**Adjectives**	**No.**	**Adjectives**
1	Global	51	Universal	101	Mega	151	Simple	201	Decent
2	Royal	52	Safe	102	Proud	152	Famous	202	Victorious
3	Best	53	Maximum	103	Positive	153	Worth	203	Warm
4	Next	54	Superior	104	Motoring	154	Continuous	204	Useful
5	Front	55	Metallic	105	Quick	155	Undisputed	205	Dead
6	Quality	56	Super	106	Middle	156	Neutral	206	Masculine
7	Very	57	Different	107	Regular	157	Mechanical	207	Unmatched
8	Standard	58	Wide	108	**Friendly**	158	Fluid	208	Forced
9	Rear	59	Hard	109	Awesome	159	Macho	209	Immediate
10	Long	60	Winning	110	Poor	160	Serious	210	Confused
11	High	61	**Sporty**	111	Slow	161	Dual	211	Loaded
12	Better	62	Nice	112	Steady	162	**Professional**	212	Favourite
13	Great	63	Environmental	113	Committed	163	Intelligent	213	Tremendous
14	National	64	Prestigious	114	Pretty	164	Incredible	214	Practical
15	Class	65	Outstanding	115	Innovative	165	Stunning	215	Satisfied
16	Excellent	66	**Heavy**	116	Regional	166	Extreme	216	Blazing
17	Latest	67	Rural	117	Exclusive	167	Narrow	217	Grey
18	Sure	68	Exciting	118	Difficult	168	Standing	218	Sufficient
19	Strong	69	Basic	119	Core	169	Focused	219	Wonderful
20	**Comfortable**	70	Blue	120	Active	170	Reliable	220	Phenomenal
21	Full	71	Stylish	121	Conventional	171	Potential	221	Splendid
22	**International**	72	Silver	122	Express	172	Asian	222	Revolutionary
23	Advanced	73	Huge	123	Thinking	173	Essential	223	Interested
24	Light	74	Running	124	Confident	174	Beat	224	English
25	Young	75	Competitive	125	Digital	175	Separate	225	Inspired
26	Genuine	76	Superb	126	American	176	Stringent	226	Associated
27	Family	77	Successful	127	Consistent	177	Rough	227	Numerous
28	Round	78	Ready	128	Multiple	178	Healthy	228	**Sharp**
29	Large	79	Happy	129	Born	179	Sudden	229	Golden
30	Easy	80	Growing	130	Convenient	180	Variable	230	Spacious
31	Economic	81	Fast	131	Modern	181	Reasonable	231	Rapid
32	Higher	82	Beautiful	132	Stable	182	Consequential	232	Classic
33	Executive	83	Sleek	133	Extensive	183	Usual	233	Exceptional
34	Special	84	Efficient	134	Remarkable	184	Bright	234	Overwhelming
35	Real	85	Chief	135	Robust	185	Critical	235	Automatic
36	Price	86	**Aggressive**	136	Integrated	186	Flying	236	Military
37	Cool	87	Extra	137	Physical	187	Crazy	237	Ordinary
38	Unique	88	Attractive	138	Dynamic	188	Plastic	238	Game
39	Black	89	Powerful	139	**Popular**	189	Brief	239	Worst
40	Important	90	Urban	140	Contemporary	190	Distinctive	240	Racy
41	Necessary	91	Choice	141	Western	191	Effective	241	Expert
42	Powered	92	Immediately	142	Interesting	192	Wrong	242	Eligible
43	Kinetic	93	Clean	143	Prime	193	Wild	243	Periodic
44	Impressive	94	Amazing	144	Sound	194	Certain	244	Touching
45	Driving	95	Automotive	145	Rich	195	Fundamental	245	Nationwide
46	Grand	96	Natural	146	Fresh	196	Common	246	Firm
47	Small	97	Moving	147	Dedicated	197	Ideal	247	Ambitious
48	Smooth	98	**Mean**	148	Dark	198	Himself	248	Advance
49	Green	99	Flash	149	Compact	199	White	249	Sustainable
50	Premium	100	Perfect	150	Idle	200	Appropriate	250	Fair

TABLE 3.3 (*Continued*)

List of 500 Motorbike-Describing Adjectives in India

	Formal		*Evocative*		*Perceptual*		*Representational*
	Historical		*Emotional*		*Behavoral*		*Broadly Evaluative*

No.	Adjectives	No.	Adjectives	No.	Adjectives	No.	Adjectives	No.	Adjectives
251	Lovely	301	British	351	Virgin	401	Youthful	451	Delighted
252	Designer	302	Excited	352	Momentous	402	Raring	452	Everlasting
253	Accurate	303	Thumping	353	Meaning	403	Glorious	453	Outdated
254	Happening	304	Surprised	354	**Formal**	404	Radiant	454	Funny
255	Substantial	305	Yellow	355	Enlightened	405	Chill	455	Striking
256	Impossible	306	Refined	356	Immense	406	European	456	Loud
257	Sizeable	307	Distinguished	357	Intensive	407	Prosperous	457	Effortless
258	Weak	308	Sensitive	358	Cheap	408	Thin	458	Countrywide
259	Directional	309	Adjustable	359	Breathtaking	409	Matching	459	Eager
260	Soft	310	Frequent	360	Acceptable	410	Concentrated	460	Uphill
261	Rare	311	Favourable	361	**Traditional**	411	Unauthorized	461	Baby
262	Evident	312	Theme	362	Afraid	412	Academic	462	Enduring
263	Vibrant	313	Disabled	363	Spirited	413	Pleased	463	Secret
264	Sensible	314	**Extraordinary**	364	Scared	414	Passionate	464	Bored
265	Hungry	315	Relentless	365	Ultimate	415	Confirmed	465	Supreme
266	Promising	316	Pleasant	366	Prospective	416	Swift	466	Inexhaustible
267	Enthusiastic	317	Relaxed	367	Careful	417	Agile	467	Unprepared
268	Lean	318	Instant	368	Typical	418	Principal	468	Fabulous
269	Travelling	319	Inevitable	369	Adult	419	Spectacular	469	Confidential
270	Engaged	320	Opposite	370	Massive	420	Restricted	470	Harsh
271	Prominent	321	Incomparable	371	Comparable	421	Dear	471	Unfamiliar
272	Utility	322	Absolute	372	Durable	422	Pilot	472	Unattended
273	Mass	323	Terrific	373	Intrinsic	423	Maroon	473	Unfair
274	Ranking	324	Affirmative	374	Lightning	424	Uncomfortable	474	Cohesive
275	Sophisticated	325	Smart	375	Honorary	425	Accidental	475	Alert
276	French	326	Comprehensive	376	Dangerous	426	Rival	476	Desirable
277	Strategic	327	Sturdy	377	Quiet	427	Brilliant	477	Interior
278	Eastern	328	Secure	378	Male	428	Absorbing	478	Inviting
279	Minimal	329	Plain	379	Peak	429	Weird	479	Tall
280	Visible	330	Glad	380	Adjacent	430	Coloured	480	Modest
281	Excess	331	Incomplete	381	Sweet	431	Orange	481	Compulsive
282	Sarcastic	332	Precise	382	Consequent	432	Rocky	482	Sculptured
283	**Emotional**	333	Promotional	383	Touched	433	Compulsory	483	Steadfast
284	Privileged	334	Ripping	384	Boring	434	Thrilled	484	Systematic
285	Sheer	335	Remote	385	Thick	435	African	485	Democratic
286	Reputed	336	Beige	386	Illustrious	436	Rugged	486	Occasional
287	Metro	337	Grayish	387	Unbelievable	437	Handsome	487	Harmful
288	Inspiring	338	Obvious	388	Worthy	438	Alive	488	Armed
289	Upcoming	339	Memorial	389	Gold	439	Dense	489	Stupid
290	Appealing	340	Congested	390	Worse	440	Enormous	490	Cooperative
291	Shaped	341	Honourable	391	Astounding	441	Trendy	491	Disappointed
292	Musical	342	Animal	392	Successive	442	Latin	492	Unusual
293	Intense	343	Celebrated	393	Passive	443	Precious	493	Dirty
294	Festive	344	Custom	394	Dominant	444	Pleasing	494	Magic
295	Empty	345	Spiral	395	Enjoyable	445	Refreshing	495	Futuristic
296	Unlike	346	Visual	396	Lazy	446	Slight	496	Noteworthy
297	Enabled	347	Pure	397	Unclaimed	447	Exacting	497	Utilitarian
298	Distinct	348	Female	398	Legendary	448	Strange	498	Rushed
299	Assured	349	German	399	Compromising	449	Animated	499	Handy
300	Fantastic	350	Mighty	400	Diverse	450	Sporting	500	Dramatic

TABLE 3.5

Frequency and Percentage Analysis (Gender-Wise)

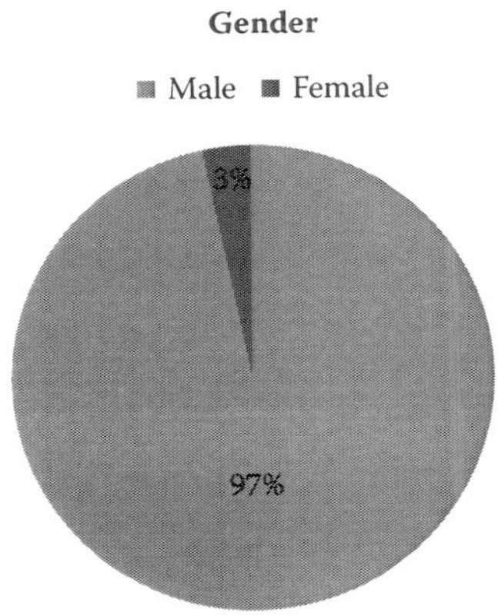

Gender	Frequency
Male	3006
Female	100
Total	**3106**

TABLE 3.6

Frequency and Percentage Analysis of Respondents in Different Age Groups

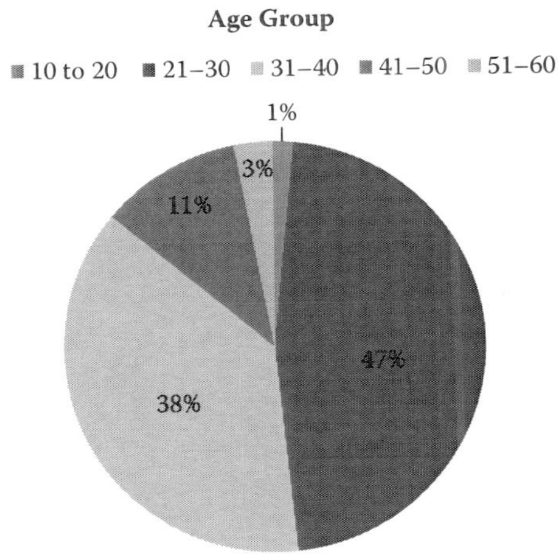

Age	Frequency
10 to 20	50
21–30	1454
31–40	1171
41–50	340
51–60	91
Total	**3106**

TABLE 3.7

Frequency and Percentage Analysis of Respondents in Relation to Occupation

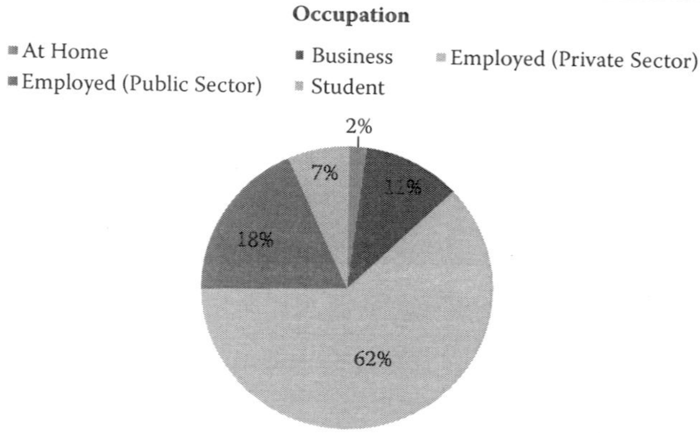

Occupation	Frequency
At Home	62
Business	334
Employed (Private Sector)	1924
Employed (Public Sector)	576
Student	210
Total	**3106**

TABLE 3.8

Frequency and Percentage Analysis of Respondents in Different Motorbike Segments

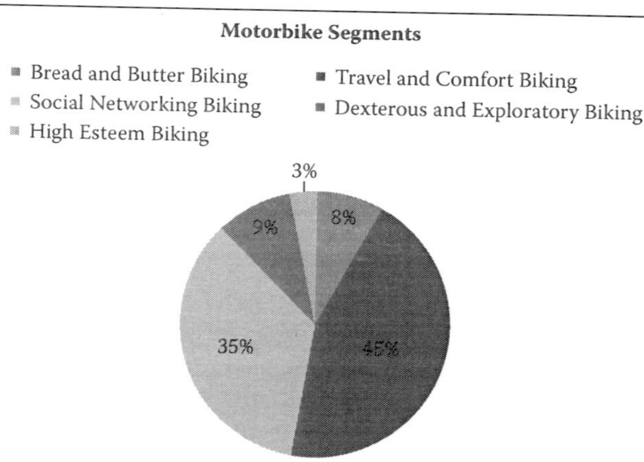

Motorbike Segments	Frequency
Bread and Butter Biking	251
Travel and Comfort Biking	1382
Social Networking Biking	1087
Dexterous and Exploratory Biking	283
High Esteem Biking	103
Total	**3106**

TABLE 3.9

Frequency and Percentage Analysis of Respondents in Relation to Their Home State

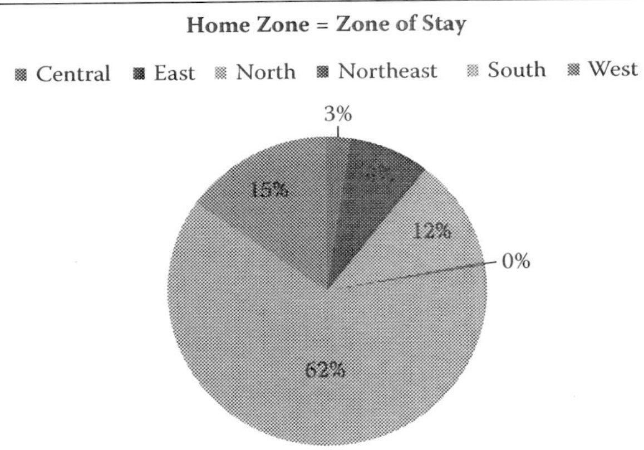

Zones	Frequency
Central	73
East	236
North	336
Northeast	10
South	1816
West	441
Total	**2912**

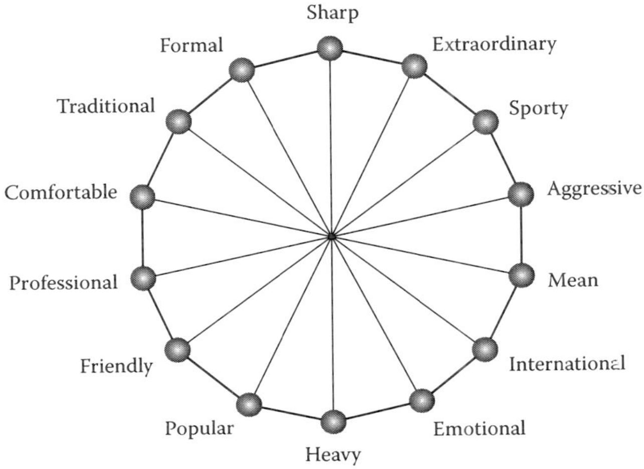

FIGURE 3.1
A cluster of 14 personality variables associated with motorbikes in India.

the 14 personality variables on a five-point rating scale (5—strongly agree, 4—agree, 3—can't say, 2—disagree, 1—strongly disagree). The instrument was subsequently used by the writer in the survey questionnaire to evaluate the biker–bike personalities of the selected sample population.

3.2 Biker–Bike Survey Questionnaire

The biker–bike personality instrument formed the basis for designing a survey questionnaire for conducting a Biker–Bike Survey with owners of motorbikes in India. A mix of closed and open-ended questions was used in the survey questionnaire, which was divided into three sections. The first section comprised questions seeking demographic information from the respondents. The biker–bike personality measurement instrument (Figure 3.2) was incorporated in the second section of the questionnaire. This section was developed in tabulated form with multiple-choice drop-down menus. The 14 personality variables (Figure 3.2) were tabulated in rows and a five-point rating scale (5—strongly agree, 4—agree, 3—can't say, 2—disagree, 1—strongly disagree) was incorporated in the drop-down menu with each personality variable. An open-ended question, "Which motorbike do you own?" was also incorporated in this section. This information was later used for segmentation of the selected sample into different biking segments on the basis of the motorbike owned. And the third section comprised an optional open-ended question, "Why do you ride a motorbike? Give three reasons."

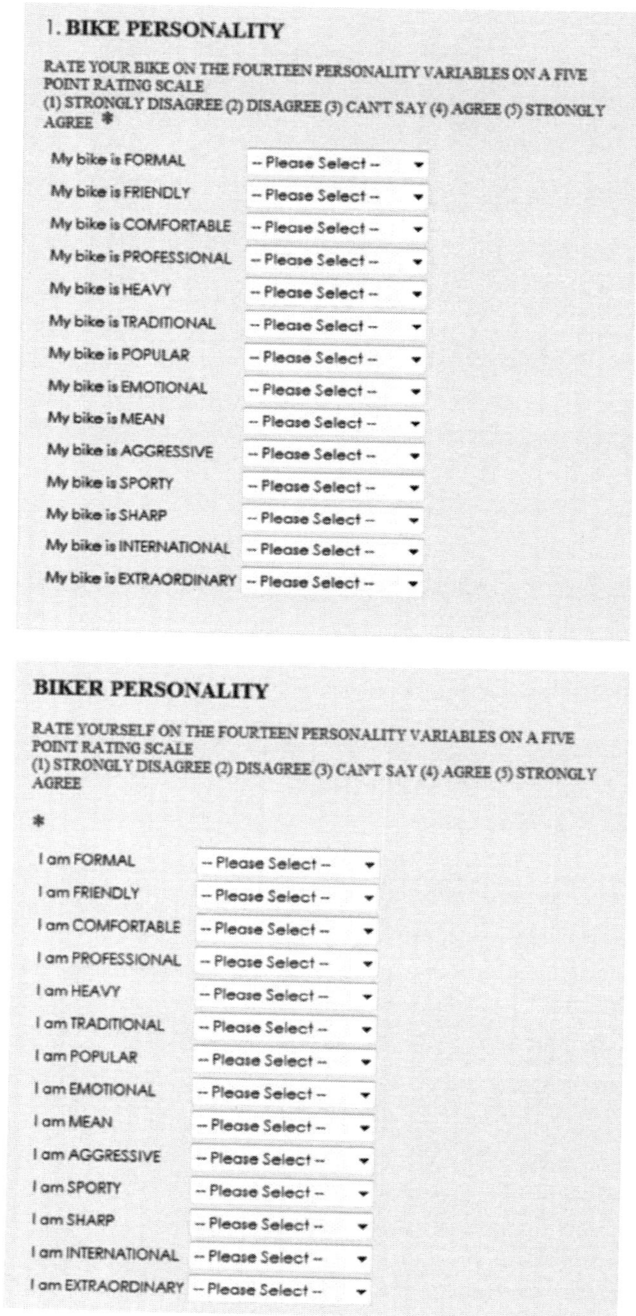

1. **BIKE PERSONALITY**

RATE YOUR BIKE ON THE FOURTEEN PERSONALITY VARIABLES ON A FIVE POINT RATING SCALE
(1) STRONGLY DISAGREE (2) DISAGREE (3) CAN'T SAY (4) AGREE (5) STRONGLY AGREE *

My bike is FORMAL	-- Please Select --
My bike is FRIENDLY	-- Please Select --
My bike is COMFORTABLE	-- Please Select --
My bike is PROFESSIONAL	-- Please Select --
My bike is HEAVY	-- Please Select --
My bike is TRADITIONAL	-- Please Select --
My bike is POPULAR	-- Please Select --
My bike is EMOTIONAL	-- Please Select --
My bike is MEAN	-- Please Select --
My bike is AGGRESSIVE	-- Please Select --
My bike is SPORTY	-- Please Select --
My bike is SHARP	-- Please Select --
My bike is INTERNATIONAL	-- Please Select --
My bike is EXTRAORDINARY	-- Please Select --

BIKER PERSONALITY

RATE YOURSELF ON THE FOURTEEN PERSONALITY VARIABLES ON A FIVE POINT RATING SCALE
(1) STRONGLY DISAGREE (2) DISAGREE (3) CAN'T SAY (4) AGREE (5) STRONGLY AGREE

*

I am FORMAL	-- Please Select --
I am FRIENDLY	-- Please Select --
I am COMFORTABLE	-- Please Select --
I am PROFESSIONAL	-- Please Select --
I am HEAVY	-- Please Select --
I am TRADITIONAL	-- Please Select --
I am POPULAR	-- Please Select --
I am EMOTIONAL	-- Please Select --
I am MEAN	-- Please Select --
I am AGGRESSIVE	-- Please Select --
I am SPORTY	-- Please Select --
I am SHARP	-- Please Select --
I am INTERNATIONAL	-- Please Select --
I am EXTRAORDINARY	-- Please Select --

FIGURE 3.2
Biker-bike personality measurement instrument.

The survey questionnaire was designed on the basis of the review of literature and opinions from experts of various related fields. It was successfully tested through a pilot study with 541 respondents before the main study.

3.3 Sampling of Bikes and Bikers

Bikers in India are not a homogeneous group. The vast geographical diversity and innumerable types of motorbikes made it extremely difficult to identify a sample population for the study. Thus, the purposive sampling method was adopted by the writer for selection of the sampling frame. The motorbike owner database of India's largest motorbike manufacturer, Hero Honda Motors Limited, was used for the study. Since the selected manufacturer has a more than 55% share in the domestic motorcycle market (Hero Honda, 2009) and a product range that covers all the proposed biking segments, the selected sample of bikes and bikers was considered most suitable for the present study of motorbikes in India.

A total of 9,512 motorbike owners belonging to 24 different states (Figure 3.3) of India constituted the sample population. The sample represented respondents from all zones (North, South, East, West, Central, and Northeast) of India. The selected sample population comprised owners of 23 different motorbikes. Figure 3.4 shows the bikes selected for the study.

Table 3.4 shows the categorization of the selected motorbikes into different motorbike segments (as explained in Section 1.5 in Chapter 1). The respondents were assigned a biking segment on the basis of the motorbikes they own.

3.3.1 Profile of Respondents

Out of the selected sample of 9,512 bikers, 553 partially completed and 3,106 fully completed the survey. The 3,106 fully complete responses were analyzed in this study. Tables 3.5–3.9 describe the respondents' profile.

A total of 3,006 male and 100 female respondents completed the survey. The respondent database comprised people from wide range of age groups. Most respondents (47%) were aged between 21 and 30 years followed by 38% aged between 32 and 40 years. Out of the total respondents, around 62% were employed in the public sector, 18% were employed in the private sector, 11% were students, and 7% were engaged in business. The remaining 2% of the sample population chose the option "at home" to describe their present work status.

The respondents were owners of 23 different Hero Honda motorbike models from different parts of the country. Depending on the motorbike owned, the respondents were categorized under different biking segments. Out of

Biker-Bike Survey

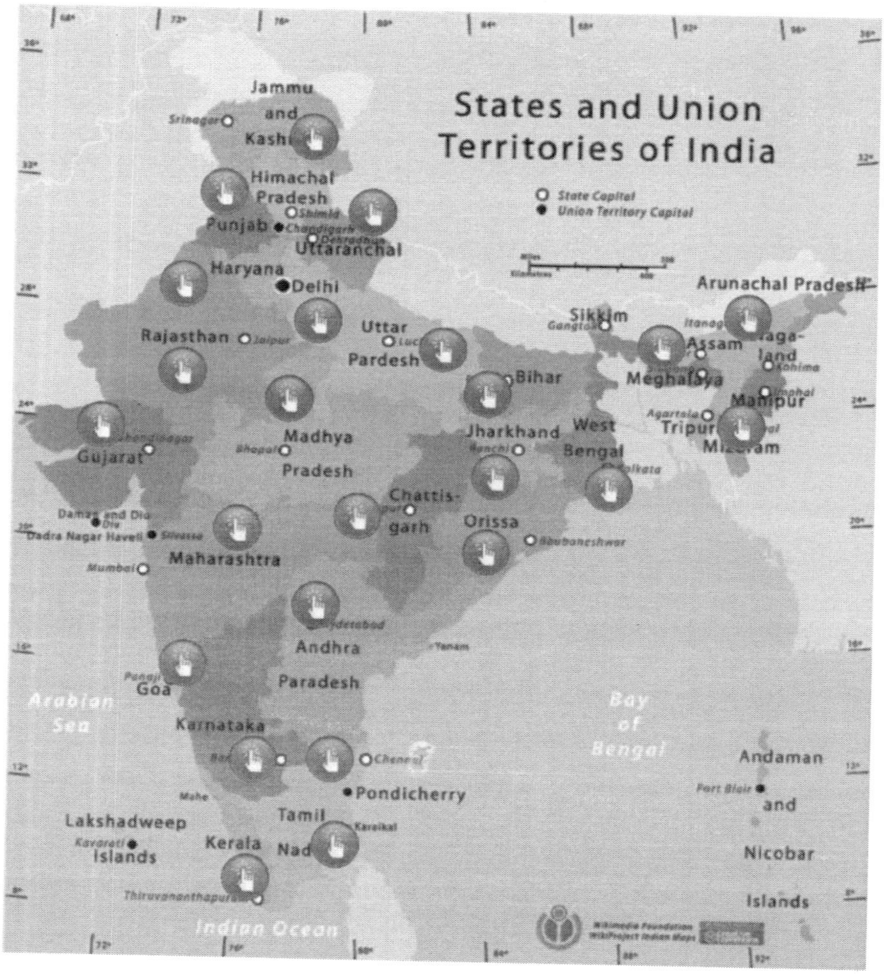

FIGURE 3.3
Statewise representation of the selected sample of 3,106 bikers.

FIGURE 3.4
Selected sample of 23 different motorbikes.

TABLE 3.4

Categorization of Selected Sample of Bikes on Different Biking Segments

Bread-and-Butter Biking	Travel-and-Comfort Biking	Social-Networking Biking	Dexterous and Exploratory Biking	High-Esteem Biking
CD DELUXE	Splendor	Passion	CBXZ	Karizma
CD 100	Splendor Plus	Passion Plus	CBZ Xtreme	Karizma R
CD 100 SS	Splendor NXG	Passion Plus Special Edition	Hunk	
CD DAWN	Super Splendor	Glamour	CBZ Star	
DAWN	Splendor Plus Special Edition	Achiever		
		Ambition		
		Ambition 135		

the overall data, 45% respondents belonged to travel-and-comfort biking, 35% to social-networking biking, 9% to dexterous and exploratory biking, 8% to bread-and-butter biking, and 3% to high-esteem biking. The representation of different biking segments is similar to the relative popularity of these biking segments in India.

The survey questionnaire asked respondents about their home state and the state where they have stayed for the maximum number of years. The

TABLE 3.5 (See color insert)

Frequency and Percentage Analysis (Gender-Wise)

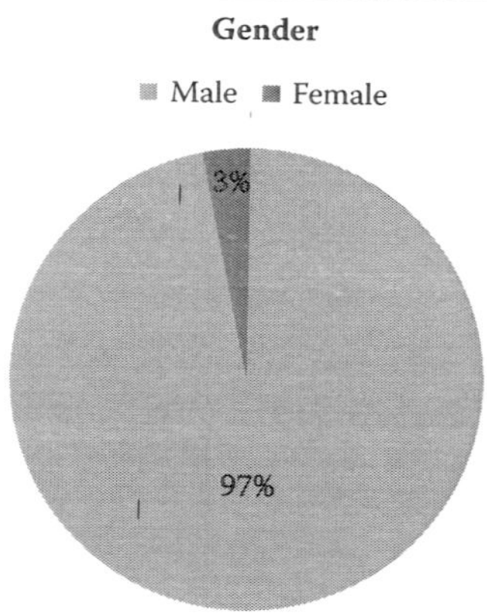

Gender	Frequency
Male	3006
Female	100
Total	**3106**

TABLE 3.6 (See color insert)

Frequency and Percentage Analysis of Respondents in Different Age Groups

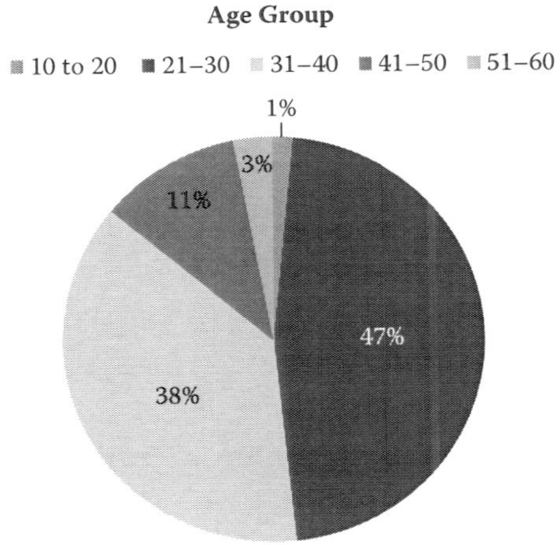

Age	Frequency
10 to 20	50
21–30	1454
31–40	1171
41–50	340
51–60	91
Total	**3106**

objective of this question was to identify those respondents who had also stayed for the maximum number of years in their home state. The objective of the question was to segregate the respondents on the basis of their cultural background. On the basis of the data recorded, it was found that 2,912 respondents had also stayed for the maximum number of years in their home state. These respondents were selected for further study. Table 3.9 shows the region-wise representation of these 2,912 participants of the survey. The percentage of participants from southern, western, northern, eastern, and central states of India was 62%, 15%, 12%, 8%, 3%, respectively. The least participation was from Northeast India, with only 10 participants responding to the survey.

TABLE 3.7 (See color insert)

Frequency and Percentage Analysis of Respondents in Relation to Occupation

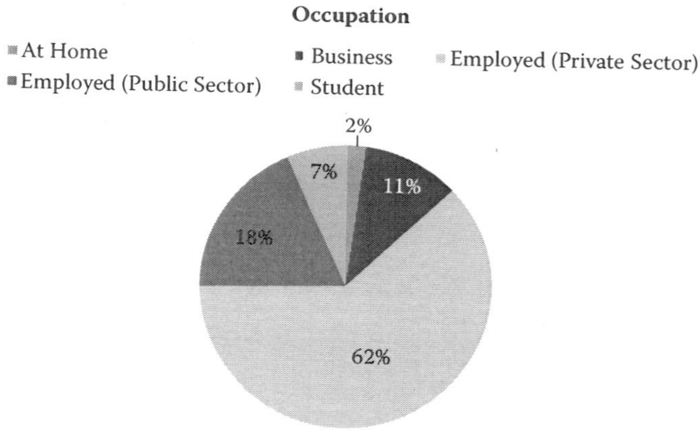

Occupation	Frequency
At Home	62
Business	334
Employed (Private Sector)	1924
Employed (Public Sector)	576
Student	210
Total	**3106**

TABLE 3.8 (See color insert)

Frequency and Percentage Analysis of Respondents in Different Motorbike Segments

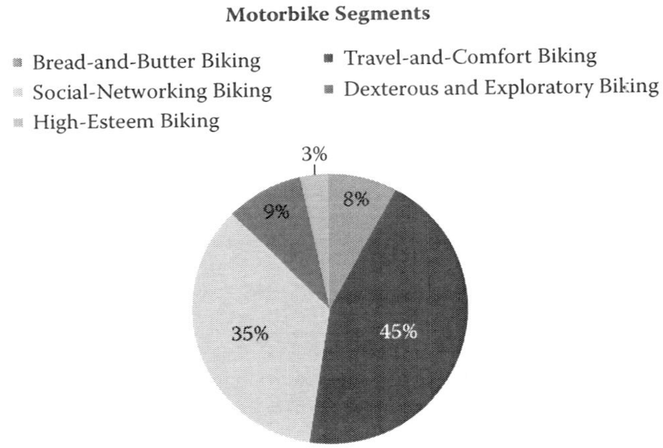

Motorbike Segments	Frequency
Bread-and-Butter Biking	251
Travel-and-Comfort Biking	1382
Social-Networking Biking	1087
Dexterous and Exploratory Biking	283
High-Esteem Biking	103
Total	**3106**

TABLE 3.9 (See color insert)

Frequency and Percentage Analysis of Respondents in Relation to Their Home State

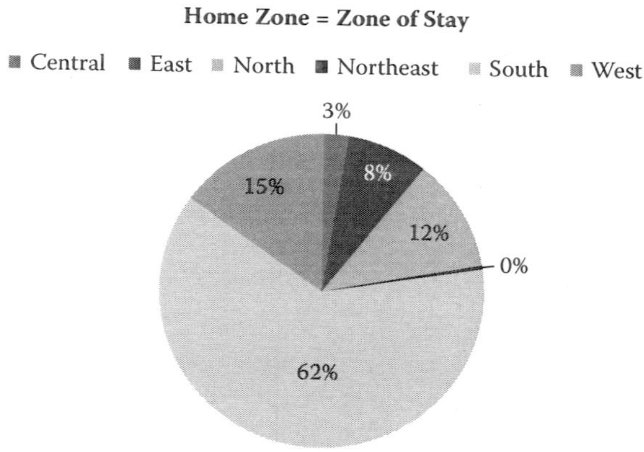

Zones	Frequency
Central	73
East	236
North	336
Northeast	10
South	1816
West	441
Total	**2912**

3.4 Data Collection Procedure

The potential respondents who were identified for the study belonged to different parts of the country. An online investigation was considered the most suitable data collection method in the context of the study. The survey questionnaire was uploaded on a registered web link on www.surveygizmo.com and circulated to the sample population by e-mail. A set of phone messages were also sent to the bikers informing them about the survey.

A pilot survey was conducted before the main study. Figures 3.5 and 3.6 show the response by location and response by day for the pilot study conducted with 541 respondents. A total of 203 respondents completed the pilot survey questionnaire, while 332 respondents abandoned the survey and 6 respondents partially completed the questionnaire. Figures 3.7 and 3.8 show the response by location and response by day for the main study with 3,106 respondents.

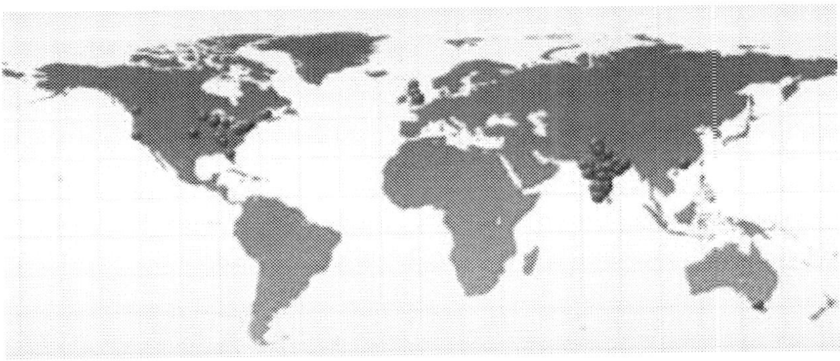

FIGURE 3.5
Response by location for the first pilot survey.

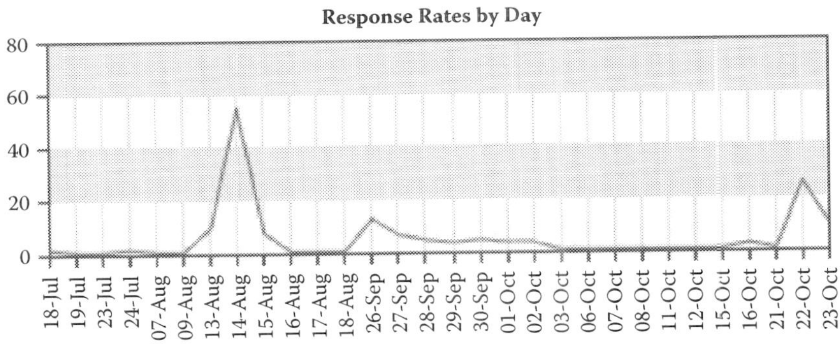

FIGURE 3.6
Response rates by day for the first pilot survey.

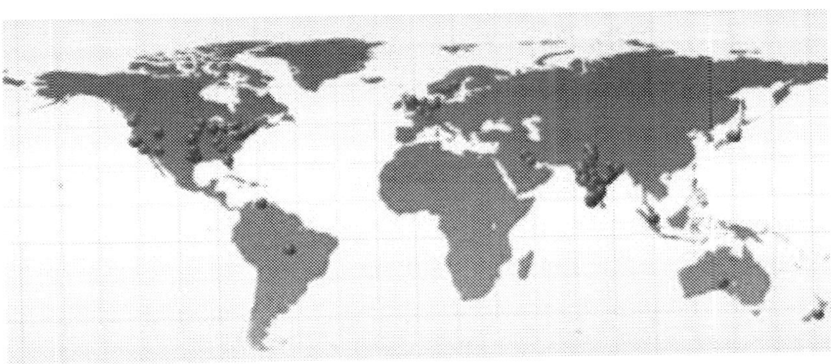

FIGURE 3.7
Response by location for the main study.

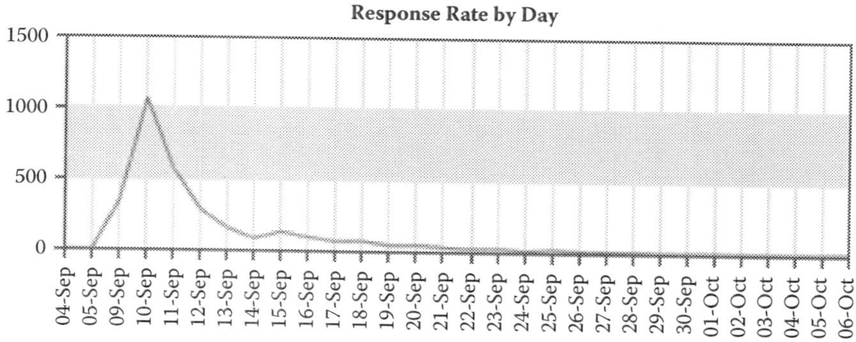

FIGURE 3.8
Response rates by day for the main study.

3.5 Techniques Used for Data Analysis

The data after collection was classified into two broad categories: qualitative and quantitative. Quantitative data included the personality ratings given by respondents on the 14 personality variables, and the qualitative information was collected from the respondents' feedback to the open-ended question. The qualitative data were analyzed first to ascertain the social-cultural segmentation of biking proposed in Chapter 1. This was followed by the quantitative analysis of the biker–bike personality ratings given by respondents to assess the emotive quality of the biking segments.

3.5.1 Qualitative Analysis: Ascertaining the Social Cultural Segmentation of Biking

The qualitative data collected for the open-ended question, "Why do you ride a motorbike? Give three reasons" was analyzed in the following steps:

1. *Identification of biking needs:* The qualitative data were coded and analyzed on the basis of a content analysis technique (Krippendorff, 2004) to categorize the reasons for biking reported by the 3,106 respondents into a set of biking needs.
2. *Segment-wise comparison of biking needs:* The biking needs related to bread-and-butter biking, travel-and-comfort biking, social-networking biking, dexterous and exploratory biking, and high-esteem biking segments were investigated on the basis of the descriptive analysis (frequency and percentage) techniques.
3. *Mapping of biking needs on H-E-A-T diagram:* The identified biking needs were mapped on the Hedonistic-Emotional-Ascetic-Technical (H-E-A-T) diagram developed by Das (2002). The objective of

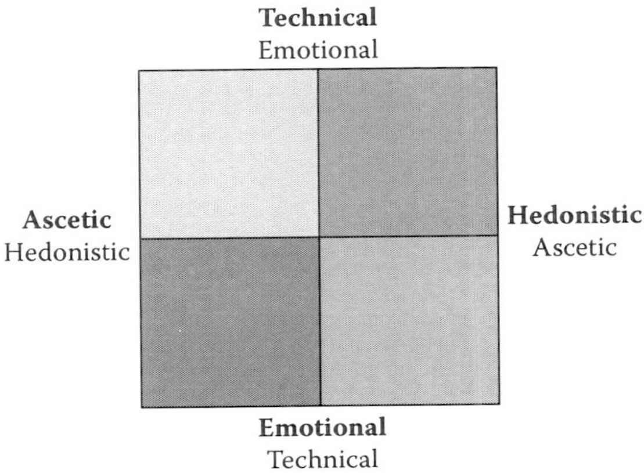

FIGURE 3.9
H-E-A-T diagram. (From Das, L. K. (2002). Toward a non-parochial, non-partisan framework for the study of design history. *Presented at the 3rd International Conference on Design History and Design Studies*, July 9–12, 2002, Istanbul, Turkey.)

mapping needs on H-E-A-T diagram was to analyze the difference in quality of needs associated with each biking segment. Figure 3.9 shows the H-E-A-T diagram. In the H-E-A-T framework, design contributions are mapped on a hedonistic-ascetic continuum on one axis and a technological-emotional continuum on the other.

4. *Mapping of affective experiences on circumplex of emotions:* The biking needs related to "affective experience" were mapped on the circumplex of emotions (Figure 2.3, Sec 2.1.4, Chapter 2) developed by Desmet (2002) to assess the difference in the intensity of emotional experience associated with each biking segment.

3.5.2 Quantitative Analysis: Emotive Quality of Biking Segments

The quantitative data reported by respondents on the biker–bike personality measurement instrument were analyzed in the following steps:

1. *Biker–Bike personality relationship*: The ratings given by respondents for their own and their bike's personality were coded and recorded in the form of tables using SPSS 13.0 software, which was then used for descriptive analysis, frequency distributions, and cross-tabulations. Cronbach's alpha test was performed to test the reliability of the survey questionnaire. The data collected was subjected to parametric tests of correlation and Pearson's chi-square test to investigate the relationship between biker–bike personalities on each personality variable.
2. *Biker and Bike personality factors*: The correlated personality variables were grouped together into personality factors using the method of

factor analysis (principal component analysis with varimax rotation technique with Kaiser Normalization).

3. Comparing biker and bike personalities of biking segments: The one-way ANOVA method was used to compare the personalities of bikers and bikes belonging to bread-and-butter biking, travel-and-comfort biking, social-networking biking, dexterous and exploratory biking, and high-esteem biking segments on the identified personality factors.
4. *Emotive quality of biking segments*: The emotive quality of each biking segment was presented in the form of a multiscatter plot diagram. The diagram represented the factor value score of each biking segment on the biker and bike personality factors.

The results of the qualitative and quantitative analysis were correlated to formulate an emotion-centered research framework for design innovation.

4

The Sociocultural Segmentation of Biking

The sociocultural segmentation of biking proposed in Chapter 1 differentiates biking in India on the basis of related biking needs. The first stage of analysis therefore is focused on establishing the significance of the proposed biking segments: *bread-and-butter biking, travel-and-comfort biking, social-networking biking, dexterous and exploratory biking,* and *high-esteem biking.*

An extensive qualitative database was generated for the question, "Why do you ride a motorbike? Give three reasons." The following sections illustrate a few examples of the responses to this question from the respondents belonging to different biking segments. The qualitative responses given to this question were comprehensively analyzed by the writer to investigate the difference in the biking needs of bikers belonging to different biking segments in India.

4.1 Bread-and-Butter Biking

Here are a few examples of answers (quoted verbatim) given to the question, "Why do you ride a motorbike?" by respondents in the category of *bread-and-butter biking*:

> A motorbike with low fuel consumption is a cheaper conveyance for me. I do not have to wait for public transport. Though it is troublesome in the rainy season, as well as in winter, it is better than four-wheelers in the city where there are crowds crossing the jammed roads.
>
> **Male, 51 years, Bihar, businessman**

> [A motorbike] can save time, and the family is happy.
>
> **Male, 45 years, Delhi, employed in public sector**

> It is convenient in respect to travel, time, etc. Less expensive, too, as my residence is very near. Gives some status in society [in contrast to using public transport].
>
> **Male, 47 years, Tamil Nadu, employed in public sector**

Quality, beauty, trustworthiness.

Male, 28 years, Delhi, employed in private sector

It is my mode of transportation for local movement. I love to ride. It is the most convenient mode of transportation for me.

Male, 34 years, Orissa, employed in private sector

It gives me comfort to reach my destination in a easy way. It is useful for all my personal and official purposes. It is economical and feasible.

Hemambar, 30 years, Andhra Pradesh, employed in private sector

For my personal use, it gives a comfortable ride and independence, A motorbike can run in many areas where four-wheelers can't. It is cheapest, too.

Male, 35 years, Orissa, employed in private sector

A motorbike is the best option for my city (Chennai) as it is swift and trendy, I can move quickly with no hassles even through traffic jams (which I encounter daily). It's economical (fuel efficient) with lower maintenance charges.

Male, 28 years, Tamil Nadu, employed in private sector

For time savings, for efficient execution of my various duties related to my profession, it is more fuel efficient than any other personal means of transport.

Male, 36 years, Kerala, employed in private sector

[I like my bike] because it is comfortable to ride on both type of roads, rough and well maintained. It is also a light motorbike.

Male, 32 years, Rajasthan, employed in public sector

4.2 Travel-and-Comfort Biking

These are a few examples of answers (quoted verbatim) given by respondents belonging to the category of *travel-and-comfort biking* to the question, "Why do you ride a motorbike?":

In Bangalore, it's a great convenience. It lets me exercise my right to freedom of movement (although this is not given by the Indian Constitution!). My wife now likes the bike rides when we go for shopping and long distances. I've become used to my bike so much that I have to say my bike is only next to my wife as the dearest thing in life now!

Male, 31 years, Karnataka, employed in public sector

It is required for daily life! It is a pleasure. It is God's gift!

Male, 31 years, Tamil Nadu, employed in private sector

I can't afford a car. I drive daily only 10 km because my office is 22 km one way and I prefer to travel to it by public transport. (I also have back problem.) Good mileage.

Male, 30 years, Tamil Nadu, employed in private sector

It is comfortable on a conventional road with its big wheels. Its suspension, its light, and horn are comfortable for me.

Male, 39 years, Tripura, employed in public sector

I like motorbike riding. I can pass anywhere even through a small lane. If I had a four-wheeler we could take broad roads only. I do not like billionaires' big cars; they would spoil my mood by diverting my attention on a joyful ride to all those family/friends who would want to come along. Every Sunday I can enjoy the city with my bike.

Male, 34 years, Tamil Nadu, employed in private sector

A motorbike is comfortable for long and constant driving. It is strong and the mileage is good. And especially Hero Honda. There's resale value only for Hero Honda because the production is very good.

Male, 26 years, Maharashtra, businessman

There's liberal mobility (without depending on another vehicle). It has cost-effectiveness, but minimum social status.

Male, 30 years, West Bengal, employed in public sector

First thing I bought with my own money from my salary. Very strong emotional attachment. Very faithful, never troubled me, and I always have a smooth and joyful ride.

Male, 30 years, Haryana, employed in private sector

Reduce backaches due to strong shock absorbers. My Hero Honda Splendor Plus gives me pleasure while riding. A very, very strong reason to use it is that it is very economical. It will save a lot of money, so what more does the middle class require? It touches the heart of middle-class people like me. *'Aam admi ki diloka sartaj.'*

Male, 44 years, Karnataka, employed in private sector

4.3 Social-Networking Biking

These are examples of answers (quoted verbatim) given by respondents belonging to *social-networking biking* to the question, "Why do you ride a motorbike?":

Easy to pass through a crowded or thick-traffic road. The best way to spend time with a girlfriend on wheels and most comfortable.

Male, Andhra Pradesh, 39 years, employed in private sector

It is more fuel efficient than any four-wheeler. It is light and sporty, and prevents a lot more traffic congestion. It suits my student's life.

Male, 23 years, Bihar, student

To go to office, to go to church, and to take my mom out.

Male, 26 years, Tamil Nadu, employed in private sector

Need it to go to office. Can give lift to parents. Can share with friends.

Male, 31 years, Pondicherry, employed in private sector

It's very easy to ride a bike in heavy traffic. With a bike, we can go anywhere, whether it's a broad way or a narrow way. While traveling on a bike, we feel free, and we can see people and other objects easily.

Male, 30 years, Kerala, employed in private sector

For fulfilling my life's dreams. To reach my goals in time. [My motorbike] empowers me and increases my confidence level.

Harendra Bipinchandra Sukhadia, 38 years, Gujrat, employed in private sector

> My bike is a very good companion of mine. He saves me time and effort … especially on the busy roads of Chennai. He never lets me feel lonely. Whenever I start to feel lonely, I'll go for a ride, taking my gf on its pillion.
>
> **Male, 24 years, Kerala, employed in private sector**

> To save time, to save money. Instead of spending huge amount for autos, taxis, etc., [I can take it] to attend parties and functions, and to meet relatives, along with friends/family members. In case of an emergency, it's ready at odd hours.
>
> **Male, 31 years, Andhra Pradesh, employed in private sector**

> Because I am a medical representative, I should have a bike. On special occasions I can take my wife to different locations and go to my mom's and mother-in-law's house along with my wife.
>
> **Male, 22 years, Andhra Pradesh, employed in private sector**

> For transportation to office and back. Chilling out with friends. To get relieved whenever I feel tired.
>
> **Male, 26 years, Kerala, employed in private sector**

4.4 Dexterous and Exploratory Biking

These are a few examples of answers (quoted verbatim) given by respondents in our category of *dexterous and exploratory biking* to the question, "Why do you ride a motorbike?":

> For the thrill. For the love of speed. To be on top.
>
> **Male, 19 years, Tamil Nadu, student**

> To look sporty. Reflects personality of the rider. Easy to ride/drive in city like Ludhiana where people have no driving sense, bumpy roads. Dents on cars are very common.
>
> **Male, 44 years, Rajasthan, employed in private sector**

> Feel like God on road. Weekend getaway and hangouts. Personal work.
>
> **Male, 21 years, Tamil Nadu, employed in private sector**

> [It's] comfortable with my personality. Flexible on Indian roads. Its style suits the road.
>
> **Harsh Mahajan, 26 years, Delhi, business**

For more pick up and better fuel consumption. For better handling and road stability. It gives me sporty pleasure and looks different from other motorbikes like Yamaha, Pulsar, TVS.

Male, 27 years, Gujarat, employed in private sector

I dreamed of being a wing man. [On this bike], I feel I'm flying high. For the sporty spirit.

Male, 28 years, Andhra Pradesh, employed in private sector

It gives me the feeling of being a free spirit. To me it symbolizes independence, flexibility, modernity, and authority. It's the most convenient way to get around the city.

Male, 22 years, Orissa, employed in private sector

I ride my bike for the thrill and enjoyment of riding it, for its easy maneuverability in traffic, and for fuel efficiency.

Male, 24 years, Karnataka, student

I love riding. I have a passion for bikes and cars. I like riding at max speed.

Male, 24 years, Karnataka, employed in private sector

I am passionate about bikes. It is one of the most convenient ways of traveling for me. The different type of amazing models Hero Honda is introducing nowadays attract me the most to ride a motorbike.

Male, 33 years, Maharashtra, employed in public sector

4.5 High-Esteem Biking

A few examples of answers (quoted verbatim) given by respondents in our category *high-esteem biking* to the question, "Why do you ride a motorbike?":

The thrill. Mean speed. I like to maneuver the bike and do some stunts.

Male, 24 years, Karnataka, employed in private sector

Bikes are my passion. Bikes are easy to ride in traffic. I love to feel the air on my body.

Male, 24 years, Rajasthan, student

Very rough, fast, and zigzags [well].

Male, 20 years, Rajasthan, student

It is my passion to ride powerful mobikes but with proper care and responsibility. The experience of driving a motorcycle is awesome, only if you feel the beauty of it and not disrespect the vehicle. I love bikes.

Male, 21 years, Haryana, student

Adrenal rush. Gives an identity. For the thrill of riding.

Male, 23 years, Delhi, student

"I love riding them. The power in a bike is just AWESOME. Apt for Bangalore traffic!!!!!!!!!

Male, 21 years, Karnataka, student

I like riding motorcycle. It gives you sense of freedom. It takes you to a different world.

Male, 34 years, Kerala, employed in private sector

I have a great passion for riding bikes and it's one of my favorite things to do … I am addicted to my bike …

Male, 21 years, Maharashtra, student

It's sexy. It's cool. It's muscular …

Male, 29 years, Karnataka, student

Had the passion for bikes as a child (grew up around them). Associate them with a sense of freedom. Love the long drives.

Male, 24 years, Andhra Pradesh, employed in private sector

4.6 Identification of Biking Needs

The survey generated more than 10,000 responses to the question, "Why do you ride a motorbike? Give three reasons." A total of 33 unique-need categories were identified by the content analysis (Krippendorff, 2004). Table 4.1 shows the identified need categories along with a few examples.

TABLE 4.1

33 Categories of Biking Needs Identified by Qualitative Content Analysis

Need Category	Code	Examples: Answers Given by Bike Users to the Question, "Why do you ride a motorbike?"	No.
Fuel efficiency	1	*I ride a motorbike as it is fuel efficient; fill it, shut it, forget it. Mileage 60 km/liter.*	647
Economy	2	*It's economical; it is within my budget—affordable, affordable price.*	591
Comfort	3	*Comfortable on journeys; comfortable in the city.*	913
Parking convenience	4	*Economical. Normally do not get stuck in traffic. Ease of parking in city.*	125
Ease in traffic	5	*In heavy traffic like in Kochi city, it is better to use a motorbike than a car.*	500
Saving time	6	*In city life a bike is a most valuable time saver. Time is everything for me.*	745
Low maintenance	7	*Maintenance is low. Reasonable spares. It's a trouble-free bike.*	215
Safety	8	*To avoid journey risks; for safety and reliability; it is very easy and safe to drive it.*	80
Looks and style	9	*My bike is sporty, my bike is sharp; it's sexy, it's cool, it's muscular; it's formal.*	419
Can't buy car	10	*I don't have a car. No money to buy one.*	31
Social status	11	*Become a status symbol. Gives some respect in society; prestige*	60
Resale value	12	*Always best in resale [value]. That's why [when I'm ready for a change, I mostly buy a new one].*	12
Attract girls	13	*My girlfriend loves it; for passion in front of girls; for impressing girls.*	40
Enhance personality	14	*Reflects personality of the rider; it builds up your personality; sense of macho.*	177
Family	15	*Comfortable travel with family; to go out with wife; to take children to classes.*	145
Commuting	16	*Office, outing, general usage; for commuting; personal use, shopping.*	553
Affective experience	17	*I love to ride; for fun; to be relaxed; pleasure, passion, enjoy riding; I like riding.*	952
"Touch the air"	18	*It feels better to be [out in the open] than to be in a box; love to feel the air while riding.*	44
Personal freedom	19	*Gives me sense of freedom; liberal mobility (without depending on other vehicles).*	114
Necessity	20	*It's a need for me; for my nature of work. It's compulsory now; necessity, urgent.*	106
Vehicle quality	21	*Quality; durable; long lasting; sturdy; robust; it's steady; dependable.*	70
Performance	22	*I feel it has more road grip than scooters; speed; good pickup; powerful.*	171
Convenience	23	*Convenient to travel; more convenient than four-wheeler; convenient mode of transport.*	232

TABLE 4.1 (*Continued*)

33 Categories of Biking Needs Identified by Qualitative Content Analysis

Need Category	Code	Examples: Answers Given by Bike Users to the Question, "Why do you ride a motorbike?"	No.
Avoid public transport	24	*To get away from the guys who hit in bus; to avoid congestion and crushing in bus*	114
Trouble in rain	25	*I like driving except for the rainy season; it is troublesome for the rainy season*	2
Accessibility	26	*For transportation purpose from one place to another place; for easy access*	340
Perform stunts	27	*For stunts, a motorbike is essential to me; very rough, fast, zigzagging; mean speed*	6
Adventure	28	*Offroading; adventure; I am a travel lover; I love riding my motorbike long distance*	45
Ease of use	29	*Easy to handle; for easier driving; easy to go around; easier to ride than a car*	344
Local travel	30	*For local convenience; fast access to short destinations; for short jobs inside cities*	50
Company brand	31	*I believe in Hero Honda brand; brand name; proud to own a Hero Honda bike*	56
Environment friendly	32	*Eco-friendly; pollution control; to save fuel for [my] country rather than wasting it with a car*	22
Popular	33	*Most popular vehicle in the world; most popular means of transport of the century*	16
		TOTAL NUMBER OF COMMENTS—7,937	

Each of the three responses to the question was separated into discrete statements, which were then assigned to one or more of these need categories. A coding system was designed to label each statement. A response was assigned more than one code if it communicated diverse needs. For example, the following comment by one of the respondents could be separated into eight statements, each representing and communicating a distinct biking need:

> It is fuel efficient (1) and within the budget (2) and there are no hassles of maintenance (3). It has a pleasing appearance (4), ease of operation (5), and also the engine is designed in such a way that the emissions are not harmful to the environment (6). It is very easy to move on the bike (7) and its speed is good, with high pickup (8).
>
> **Rohit Srivastava, 32 years, Jharkhand, employed in private sector, owner of Splendor Plus motorbike**

Assigned needs are (1) fuel efficiency, (2) economy, (3) low maintenance, (4) looks and style, (5) ease of use, (6) environment friendly, (7) accessibility, and (8) performance

On the other hand, a response was assigned a single code if it could be broken into discrete statements all representing the same need. For example, the following response was assigned a single code, as each of its three discrete statements represented the same need:

> My bike is sporty (1), my bike is sharp (2), my bike is formal (3).
>
> **Kuldeep Tyagi, 29 years, Uttar Pradesh, employed in private sector, owner of Glamour motorbike**

Assigned needs are (1), (2), and (3) looks and style.

A total of 7,937 statements were categorized into the identified 33 needs. Table 4.2 shows the overall and segment-wise frequency of statements that were assigned to each need category. And Table 4.3 shows the overall and segment-wise percentage of statements that were assigned to each need category.

Figure 4.1 shows the radar chart that represents the percentage of statements assigned to each need category by the overall sample. It is clear from Figure 4.1 that the need for an "affective experience" is the key need associated with biking followed by the needs for "comfort," "saving time," "fuel efficiency," and "economy."

The finding of "affective experience" as the most popular need for biking was a significant contribution to this study, as it highlights the importance of emotions in this study.

4.7 Segment-Wise Comparison of Biking Needs

In the second stage, the biking needs of the sample population were analyzed with respect to each biking segment. Figure 4.1 shows the radar chart representing the relative importance of each need category of the five biking segments.

The key segment-wise biking needs are as follows:

Bread-and-butter biking: Fuel efficiency, economy, comfort, low maintenance, necessity

Travel-and-comfort biking: Comfort, affective experience, saving time, fuel efficiency, economy

Social-networking biking: Affective experience, comfort, saving time, commuting, accessibility

Dexterous and exploratory biking: Affective experience, looks and style, personality enhancement

High-esteem biking: Affective experience, influence on personality, performance, looks and style, personal freedom, "touch the air"

Figure 4.2 illustrates the differentiation in the biking needs associated with each biking segment. *Bread-and-butter biking* is associated with needs for "fuel

TABLE 4.2

Frequency of Statements Assigned to Each Need Category

Study No.	Need Category	Code Name	Code	Total	BBB	TCB	SNB	DEB	HEB
1	Affective experience	EMO	17	952	50	365	320	156	61
2	Comfort	COM	3	913	90	429	327	51	16
3	Saving time	TIM	6	745	66	355	253	60	11
4	Fuel efficiency	FUE	1	647	82	339	191	28	7
5	Economy	ECO	2	591	79	285	196	26	5
6	Commuting	CMM	16	553	34	249	216	44	10
7	Ease in traffic	TRA	5	500	45	220	173	45	17
8	Looks and style	LOO	9	419	33	157	159	53	17
9	Ease of use	EAS	29	344	34	175	109	17	9
10	Accessibility	ACC	26	340	26	130	147	29	8
11	Convenience	CON	23	232	22	105	79	23	3
12	Low maintenance	MAI	7	215	28	115	65	5	2
13	Enhance personality	PER	14	177	7	57	51	42	20
14	Performance	PER	22	171	16	55	54	30	16
15	Family	FAM	15	145	15	82	44	3	1
16	Parking convenience	PAR	4	125	13	58	45	5	4
17	Personal freedom	FRE	19	114	7	43	43	12	9
18	Avoid public transport	PUB	24	114	8	44	53	8	1
19	Necessity	NEC	20	106	13	47	34	11	1
20	Safety	SAF	8	80	9	41	24	6	0
21	Vehicle quality	QUA	21	70	7	37	22	3	1
22	Social status	STA	11	60	5	32	16	6	1
23	Company brand	BRA	31	56	4	30	20	1	1
24	Local travel	LOC	30	50	5	22	17	5	1
25	Adventure	ADV	28	45	2	19	10	8	6
26	Touch the air	AIR	18	44	0	8	17	10	9
27	Attract girls	GIR	13	40	1	11	13	10	5
28	Can't buy car	CAR	10	31	5	15	9	2	0
29	Environment friendly	ENN	32	22	1	11	9	1	0
30	Popular	POP	33	16	3	8	5	0	0
31	Resale value	RES	12	12	1	5	6	0	0
32	Perform stunts	STU	27	6	1	0	1	2	2
33	Trouble in rain	RAI	25	2	1	1	0	0	0
			Total	7,937	713	3,550	2,728	702	244

Note: BBB: bread-and-butter biking; TCB: travel-and-comfort biking; SNB: social-networking biking; DEB: dexterous and exploratory biking; HE3: high-esteem biking.

TABLE 4.3
Percentage of Statements Assigned to Each Need

Study No.	Need Category	Code Name	Code	Total	BBB	TCB	SNB	DEB	HEB
1	Affective experience	EMO	17	11.99	7.01	10.28	11.73	22.22	25.00
2	Comfort	COM	3	11.50	12.62	12.08	11.99	7.26	6.56
3	Saving time	TIM	6	9.39	9.26	10.00	9.27	8.55	4.51
4	Fuel efficiency	FUE	1	8.15	11.50	9.55	7.00	3.99	2.87
5	Economy	ECO	2	7.45	11.08	8.03	7.18	3.70	2.05
6	Commuting	CMM	16	6.97	4.77	7.01	7.92	6.27	4.10
7	Ease in traffic	TRA	5	6.30	6.31	6.20	6.34	6.41	6.97
8	Looks and style	LOO	9	5.28	4.63	4.42	5.83	7.55	6.97
9	Ease of use	EAS	29	4.33	4.77	4.93	4.00	2.42	3.69
10	Accessibility	ACC	26	4.28	3.65	3.66	5.39	4.13	3.28
11	Convenience	CON	23	2.92	3.09	2.96	2.90	3.28	1.23
12	Low maintenance	MAI	7	2.71	3.93	3.24	2.38	0.71	0.82
13	Enhance personality	PER	14	2.23	0.98	1.61	1.87	5.98	8.20
14	Performance	PER	22	2.15	2.24	1.55	1.98	4.27	6.56
15	Family	FAM	15	1.83	2.10	2.31	1.61	0.43	0.41
16	Parking convenience	PAR	4	1.57	1.82	1.63	1.65	0.71	1.64
17	Personal freedom	FRE	19	1.44	0.98	1.21	1.58	1.71	3.69
18	Avoid public transport	PUB	24	1.44	1.12	1.24	1.94	1.14	0.41
19	Necessity	NEC	20	1.34	1.82	1.32	1.25	1.57	0.41
20	Safety	SAF	8	1.01	1.26	1.15	0.88	0.85	0.00
21	Vehicle quality	QUA	21	0.88	0.98	1.04	0.81	0.43	0.41
22	Social status	STA	11	0.76	0.70	0.90	0.59	0.85	0.41
23	Company brand	BRA	31	0.71	0.56	0.85	0.73	0.14	0.14
24	Local travel	LOC	30	0.63	0.70	0.62	0.62	0.71	0.41
25	Adventure	ADV	28	0.57	0.28	0.54	0.37	1.14	2.46
26	"Touch the air"	AIR	18	0.55	0.00	0.23	0.62	1.42	3.69
27	Attract girls	GIR	13	0.50	0.14	0.31	0.48	1.42	2.05
28	Can't buy car	CAR	10	0.39	0.70	0.42	0.33	0.28	0.00
29	Environment friendly	ENN	32	0.28	0.14	0.31	0.33	0.14	0.00
30	Popular	POP	33	0.20	0.42	0.23	0.18	0.00	0.00
31	Resale value	RES	12	0.15	0.14	0.14	0.22	0.00	0.00
32	Perform stunts	STU	27	0.08	0.14	0.00	0.04	0.28	0.82
33	Trouble in rain	RAI	25	0.03	0.14	0.03	0.00	0.00	0.00
			Total	100.00	100.00	100.00	100.00	100.00	100.00

Note: BBB: bread-and-butter biking; TCB: travel-and-comfort biking; SNB: social networking-biking; DEB: dexterous and exploratory biking; HEB: high-esteem biking.

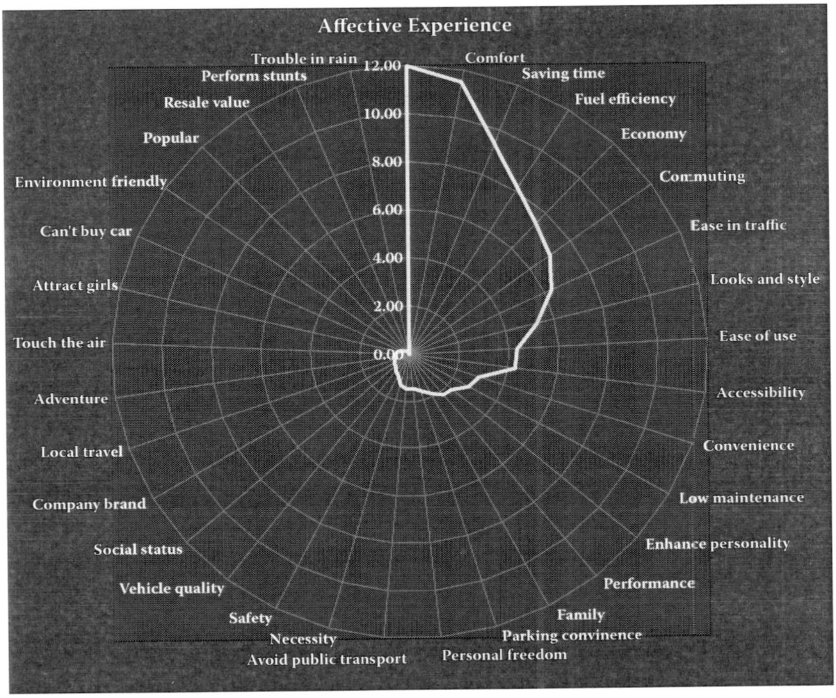

FIGURE 4.1
Radar chart representation of percentage of statements assigned to each need category.

efficiency," "economy" and "low maintenance." Biking is a necessity rather than a choice for these customers. The rational needs like "fuel efficiency" and "economy" dominate over the need for an "affective experience." On the other hand, "comfort," "affective experience," and "saving time" represent the needs for *travel and comfort biking*. Greater importance of the needs of "accessibility" and "commuting" differentiate a *social-networking biking* from *travel-and-comfort biking*. For bikers belonging to *dexterous and exploratory biking* as well as *high-esteem biking* segments, biking is primarily an emotive experience. While "looks and style" is the key need for *dexterous and exploratory biking*, the need to "enhance personality" and to acquire "personal freedom" is important for *high-esteem biking*.

4.8 Mapping of Biking Needs on H-E-A-T Diagram

The previous section differentiated the five biking segments on the basis of the relative importance of 33 different need categories. The initial review

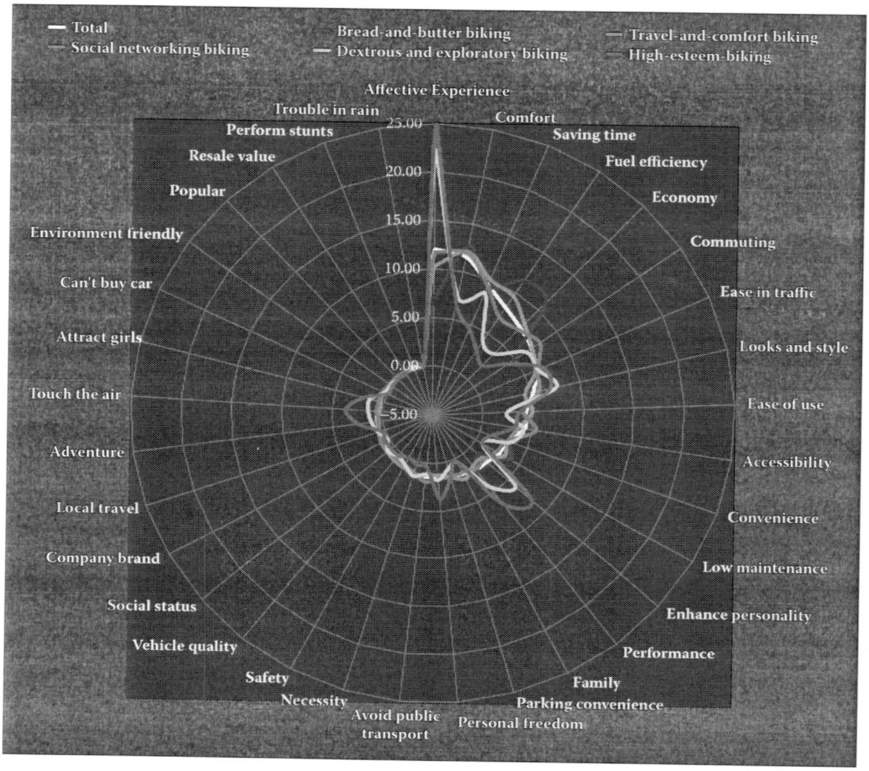

FIGURE 4.2 (See color insert)
Comparison between biking needs for each biking segment.

of the need categories suggested that though some needs are personal to the participants, there are others that represent the collective concerns of people. Also it was observed that the needs could be divided into two categories: emotional and technical. In order to understand the social significance and emotional–technical quality of various needs, it was decided to further categorize these needs on the Hedonistic–Emotional–Ascetic–Technical (H-E-A-T) diagram developed by Das (2002). In the H-E-A-T diagram, design contributions are mapped on the hedonistic–ascetic continuum on one axis and technological–emotional continuum on the other.

Figure 4.3 shows the mapping of all identified 33 biking needs on the H-E-A-T diagram. The H-E-A-T diagram divides the biking needs into four categories: hedonistic–technical needs, hedonistic–emotional needs, emotional–ascetic needs, and ascetic–technical needs. Most of the needs in the technical–hedonistic region, such as "ease in traffic," "comfort," "ease of use," "parking convenience," "vehicle quality," and so forth, are common to most bikers in India. Similar is the case with needs such as "fuel efficiency," "economy," and "low maintenance," which lie in the technical–ascetic region.

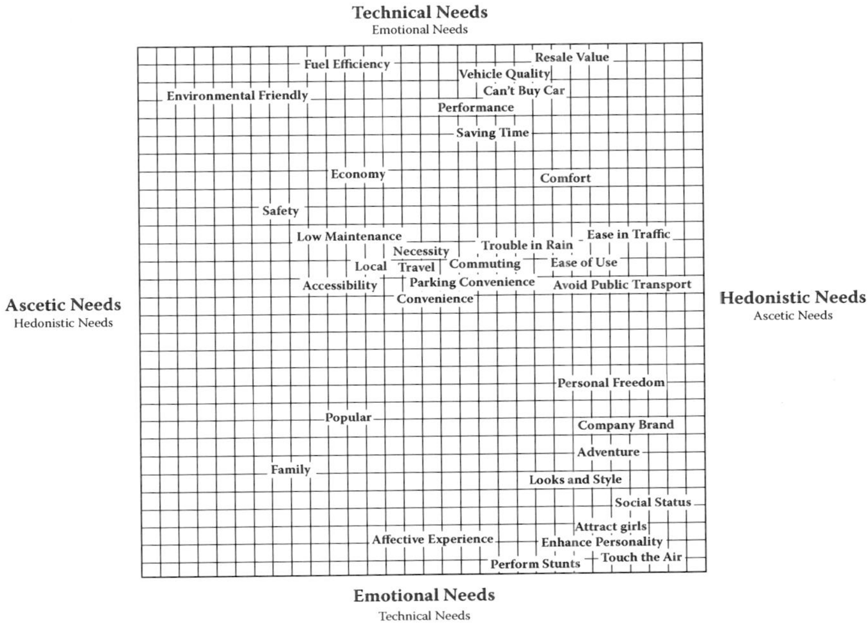

FIGURE 4.3
Mapping of biking needs on the H-E-A-T diagram.

On the other hand, hedonistic–emotional needs are more personal in nature and therefore, may not be similar for each individual. For example, need for "looks and style," "personal freedom," "social status," and "enhance personality" represent individual aspirations and preferences.

Table 4.4 and Figure 4.4 show the total percentage of hedonistic–technical needs, hedonistic–emotional needs, emotional–ascetic needs and ascetic–technical needs associated with different biking segments.

Figure 4.4 shows the differentiation in the biking needs associated with each biking segment. The percentage of ascetic–emotional needs, of biking are highest for *bread-and-butter biking* (2.52%), followed by *travel-and-comfort biking* (2.54%), *social-networking biking* (1.80%), *dexterous and exploratory biking* (0.43%), and least for *high-esteem biking* (0.41%). A similar trend is seen for ascetic–hedonistic needs, which are highest for *bread-and-butter biking* (31.56%) followed by *travel-and-comfort biking* (25.94%), *social-networking biking* (23.17%), *dexterous and exploratory biking* (13.35%), and least for *high-esteem biking* (9.02%).

The hedonistic-technical needs too follow the similar trend. However, the hedonistic–emotional needs follow a reverse trend. The highest percentage of hedonistic–emotional needs are associated with *high-esteem biking* (53.69) followed by *dexterous and exploratory biking* (42.74%), *social-networking biking* (23.83%), *travel-and-comfort biking* (20.34%), and least for *bread-and-butter biking*

TABLE 4.4

Percentage of H-E-A-T Needs Associated with Different Biking Segments

	Total	BBB	TCB	SNB	DEB	HEB
Ascetic–Emotional						
Family	1.83	2.10	2.31	1.61	0.43	0.41
Popular	0.20	0.42	0.23	0.18	0.00	0.00
Total	**2.03**	**2.52**	**2.54**	**1.80**	**0.43**	**0.41**
Ascetic–Technical						
Fuel efficiency	8.15	11.50	9.55	7.00	3.99	2.87
Economy	7.45	11.08	8.03	7.18	3.70	2.05
Accessibility	4.28	3.65	3.66	5.39	4.13	3.28
Low maintenance	2.71	3.93	3.24	2.38	0.71	0.82
Safety	1.01	1.26	1.15	0.88	0.85	0.00
Environment friendly	0.28	0.14	0.31	0.33	0.14	0.00
Total	**23.88**	**31.56**	**25.94**	**23.17**	**13.53**	**9.02**
Hedonistic–Emotional						
Affective experience	11.99	7.01	10.28	11.73	22.22	25.00
Looks and style	5.28	4.63	4.42	5.83	7.55	6.97
Enhance personality	2.23	0.98	1.61	1.87	5.98	8.20
Personal freedom	1.44	0.98	1.21	1.58	1.71	3.69
Social status	0.76	0.70	0.90	0.59	0.85	0.41
Company brand	0.71	0.56	0.85	0.73	0.14	0.41
Adventure	0.57	0.28	0.54	0.37	1.14	2.46
"Touch the air"	0.55	0.00	0.23	0.62	1.42	3.69
Attract girls	0.50	0.14	0.31	0.48	1.42	2.05
Perform stunts	0.08	0.14	0.00	0.04	0.28	0.82
Total	**24.10**	**15.43**	**20.34**	**23.83**	**42.74**	**53.69**
Hedonistic–Technical						
Comfort	11.50	12.62	12.08	11.99	7.26	6.56
Save time	9.39	9.26	10.00	9.27	8.55	4.51
Commuting	6.97	4.77	7.01	7.92	6.27	4.10
Ease in traffic	6.30	6.31	6.20	6.34	6.41	6.97
Ease of use	4.33	4.77	4.93	4.00	2.42	3.69
Convenience	2.92	3.09	2.96	2.90	3.28	1.23
Performance	2.15	2.24	1.55	1.98	4.27	6.56
Parking convenience	1.57	1.82	1.63	1.65	0.71	1.64
Avoid public transport	1.44	1.12	1.24	1.94	1.14	0.41
Necessity	1.34	1.82	1.32	1.25	1.57	0.41
Vehicle quality	0.88	0.98	1.04	0.81	0.43	0.41
Local travel	0.63	0.70	0.62	0.62	0.71	0.41
Can't buy car	0.39	0.70	0.42	0.33	0.28	0.00

TABLE 4.4 (*Continued*)

Percentage of H-E-A-T Needs Associated with Different Biking Segments

	Total	BBB	TCB	SNB	DEB	HEB
Resale value	0.15	0.14	0.14	0.22	0.00	0.00
Trouble in rain	0.03	0.14	0.03	0.00	0.00	0.00
Total	**49.99**	**50.49**	**51.18**	**51.21**	**43.30**	**36.89**

Note: BBB: bread-and-butter biking; TCB: travel-and-comfort biking; SNB: social-networking biking; DEB: dexterous and exploratory biking; HEB: high-esteem biking.

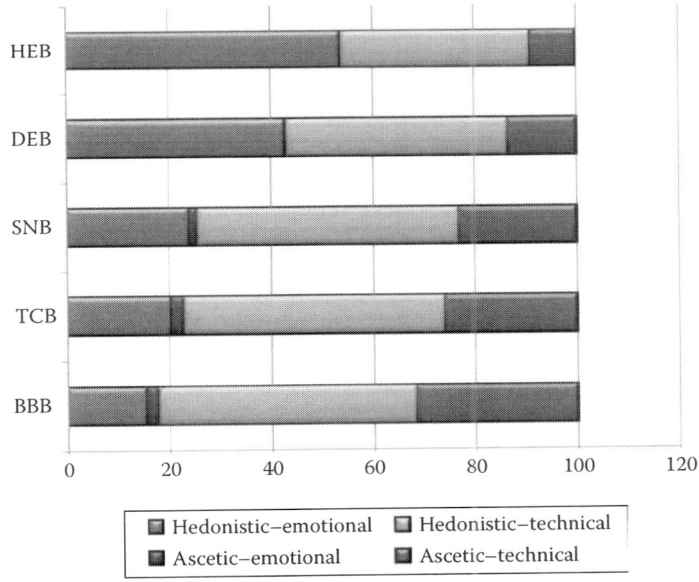

FIGURE 4.4 (See color insert)
Percentage of H-E-A-T needs associated with different biking segments.

(15.43%). As is clear from Figure 4.8, the importance of hedonistic–emotional needs increases as we move from *bread-and-butter biking* to *high-esteem biking*. However, a reverse trend is seen for hedonistic–technical needs, emotional–ascetic needs, and ascetic–technical needs, which decrease in importance from *bread-and-butter biking* to *high-esteem biking*.

Figure 4.5 further highlights this differentiation on the basis of technical and emotional needs. It is seen that the importance of technical needs decreases and emotional needs increases as we move from *bread-and-butter biking* to *high-esteem biking*. Figures 4.6–4.9 compare the significance of hedonistic–technical needs, hedonistic–emotional needs, emotional–ascetic needs, and ascetic–technical needs for different biking segments.

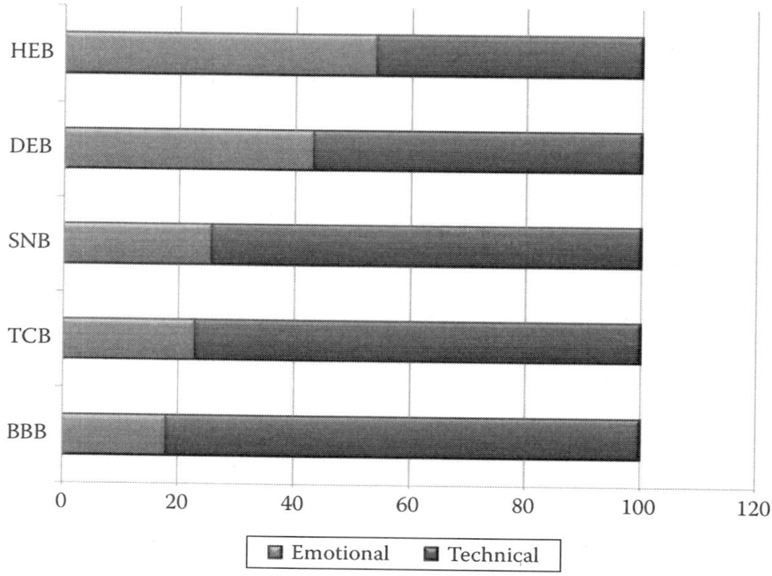

FIGURE 4.5 (See color insert)
Comparison of biking needs associated with different biking segments on the emotional–technical dimensions.

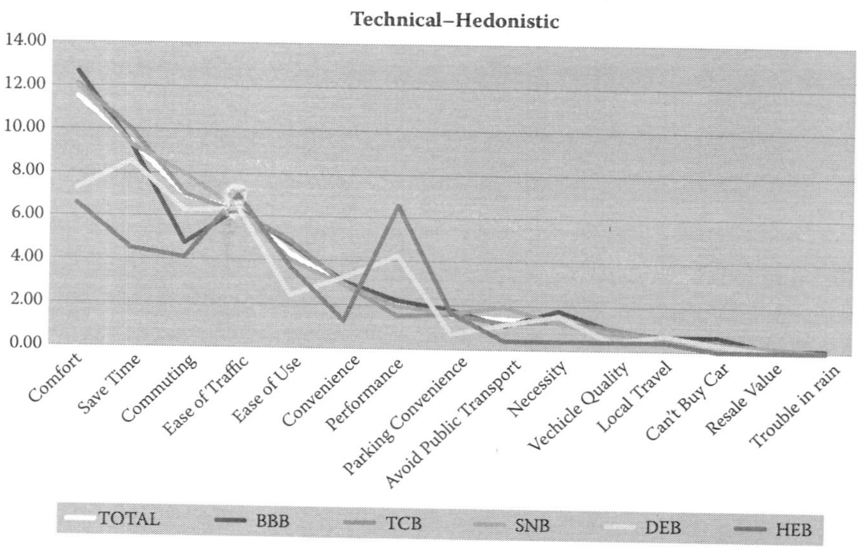

FIGURE 4.6 (See color insert)
Technical–hedonistic biking needs associated with different biking segments.

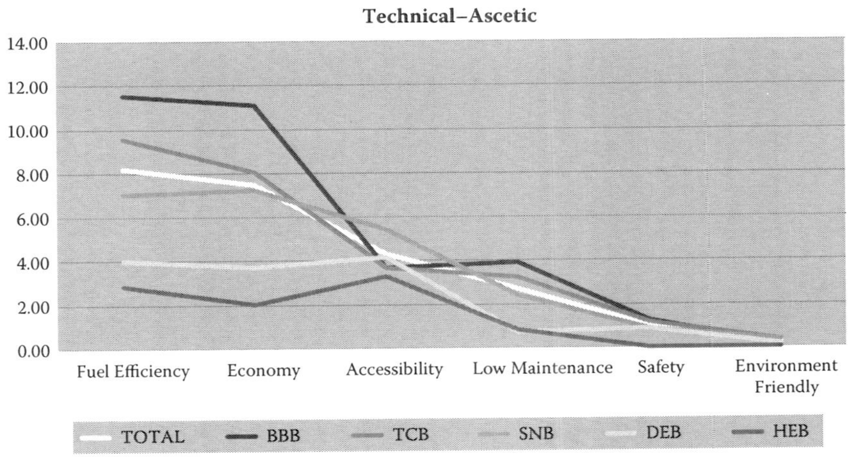

FIGURE 4.7 (See color insert)
Technical–ascetic biking needs associated with different biking segments.

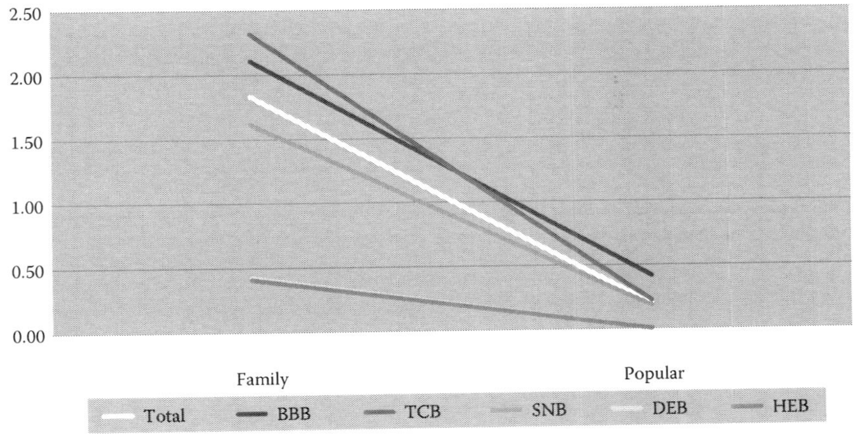

FIGURE 4.8 (See color insert)
Emotional–ascetic biking needs associated with different biking segments.

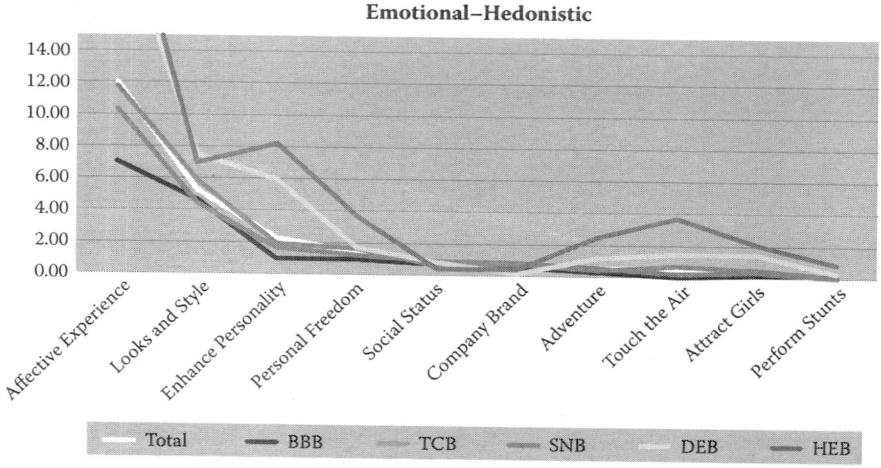

FIGURE 4.9 (See color insert)
Emotional–hedonistic biking needs associated with different biking segments.

4.9 Mapping of Affective Experiences on the Circumplex of Emotions

The findings of Section 4.1 emphasized the importance of "affective experience" as the most significant biking need of the sample under study. In the last stage of this qualitative analysis, the need for "affective experience" was further analyzed to investigate the intensity of affective experience related to each biking segment.

The 647 statements that were previously coded within the need category of "affective experience" were analyzed to identify keywords that communicate the affective state of the biker. The following unique words could be identified: *thrill, ultimate, awesome, craze, enthusiastic, ecstatic, adrenaline rush, passion, chill, dream, ambition, achieve, love, heart, like my wife or friend, part of my life, enjoy, joyful, jolly, joy, pleasure, pleasant, fun, happy, wonderful, mazaa (Hindi word meaning joy), feel good, liking, hobby, interest, keen, time pass, motivation, satisfaction, relaxation*, and *peace of mind*.

These 36 words that signify the user's affective experience were further mapped onto the low activation–high activation and pleasant–unpleasantness dimension of the circumplex of product emotions developed by Russell (1980) and Watson and Tellgen (1985) and adapted by Desmet (2002), as shown in Figure 4.10. The words could be categorized into six clusters that were named after one of the associated words as (1) *relax*, (2) *interest*, (3) *joy*, (4) *love*, (5) *passion*, and (6) *thrill* (Figure 4.10). The identified clusters were primarily differentiated on the intensity (low activation–high activation) dimension.

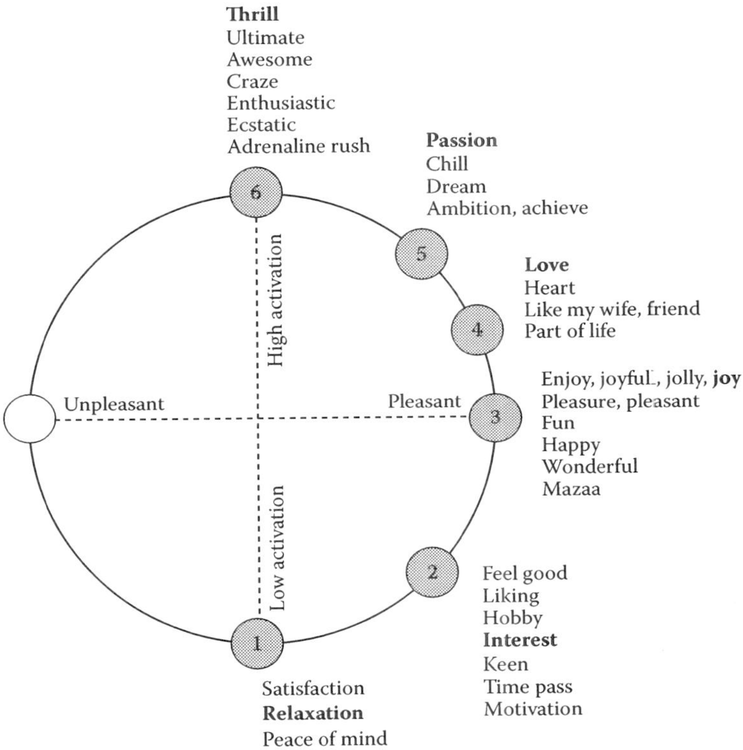

FIGURE 4.10
Mapping of words related to affective experience on the circumplex of emotions (adapted from Desmet, 2002).

Similar to Section 4.1, content analysis technique was used to code the statements within the category of "affective experience" on the six identified clusters. Figure 4.11 illustrates the differentiation in biking segments on the basis of the intensity of associated affective experience.

The intensity of affective experiences increases from left (*relax*) to right (*thrill*) in Figure 4.11. It is clear from the figure that *love* is the most popular affective experience followed by *passion, enjoy, interest, relax,* and *thrill.* However, the individual biking segments tell a different story. Notable different experiences are associated with *bread-and-butter biking* and *travel-and-comfort biking.* The former is associated with low-intensity pleasant emotions like *interest* and *enjoy* unlike the latter, where *love* and *passion* represents the key emotions. The emotion of *love* clearly represents the affective experience of bikers belonging to the *social-networking biking* with more than 65% of the statements falling under this cluster of words. The emotions associated with *dexterous and exploratory biking* and *high-esteem biking* are more diverse as compared to other segments. The former is associated with emotions such

BBB: Bread and Butter Biking
TCB: Travel and Comfort Biking
SNB: Social Networking Biking
DEB: Dextrous and Exploratory Biking
HEB: High Esteem Biking

	Total	BBB	TCB	SNB	DEB	HEB
Relax	2.13	2.13	0.56	1.25	8.50	0.00
Interest	7.47	21.28	3.07	5.64	17.65	6.67
Enjoy	6.94	38.30	2.23	4.39	10.46	15.00
Love	55.28	29.79	56.42	65.20	46.41	38.33
Passion	27.21	4.26	37.71	23.51	16.99	28.33
Thrill	0.96	4.26	0.00	0.00	0.00	11.67

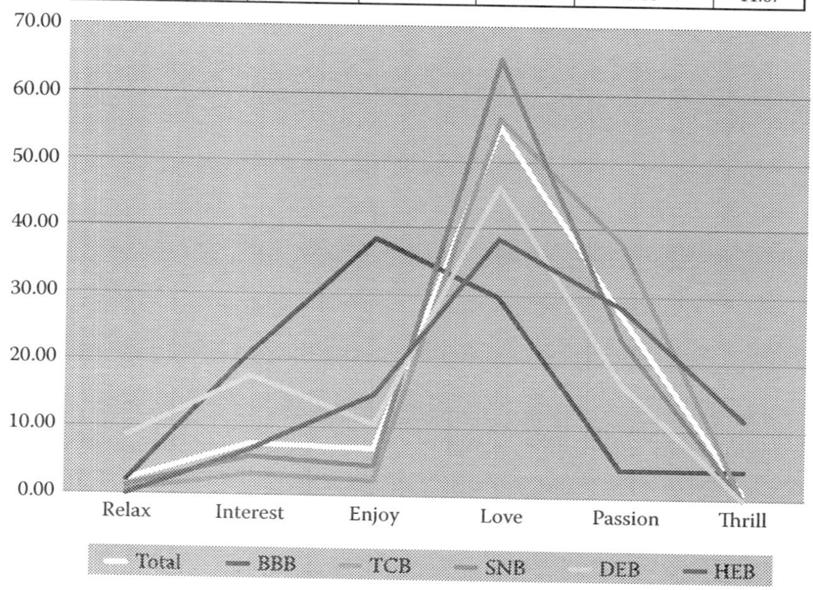

FIGURE 4.11 (See color insert)
Affective experience related to each biking segment.

as *relax, interest, enjoy, love, and passion*. On the other hand, *high-esteem biking* experiences range between *enjoy, love, passion,* and *thrill.*

The findings thereby establish a clear differentiation among the five biking levels based on the intensity of affective experiences.

4.10 Discussion

Language is a carrier of emotions. People often express their emotions and feelings in verbal or written form. The use of the open-ended question in survey questionnaire proved to be a useful strategy to record the emotional needs, aspirations, beliefs, and expectations of bikers from their bikes. The exceptional response to the survey (35.5% respondents completed the survey out of a total of 9,512 selected sample) and the fact that most of the respondents answered to the open-ended optional question indicates the emotional attachment and commitment of bikers toward their bikes. The results of the content analysis establish that the need for an affective experience is the most popular need of bikers in India.

The results of this stage of the study establish the difference in the configuration of biking needs associated with each biking segment. The results therefore ascertain the social-cultural segmentation of biking in India.

5

Emotive Quality of Biking Segments

This chapter presents the statistical analysis of the personality ratings given by respondents on the biker–bike personality measurement instrument.

5.1 Biker–Bike Personality Relationship

The writer used Pearson's chi-square test to the responses given by respondents for rating their own personality and their bike's personality to further establish the study. The primary objective was to study the relationship between the biker–bike personalities for each personality variable. The secondary objective was to identify the presence of a pattern, if any, in which the overall sample population, rates itself and its bikes on the various personality variables.

The Pearson's chi-square significance value was examined for each personality variable to study the presence or absence of a distinct relationship between the biker–bike personalities. A cross tabulation was then generated for each personality variable to study the rating (5—strongly agree, 4—agree, 3—can't say, 2—disagree, 1—strongly disagree) given by respondents for their own personality and that of their bike's. The cross-tabulation table was further analyzed to study the presence of a pattern if any in which the respondents rated their own personality with respect to the rating given to their bikes. And last, the findings of the overall 14 personality variables were compared and analyzed to explain the biker–bike personality relationship for the overall sample population.

Tables 5.1–5.14 show the results of the cross tabulations and Pearson's chi-square test conducted for each personality variable. The results show a clear and significant relationship (significance value of 0 for 16 degrees of freedom) between the biker and bike personality for each of the 14 personality variables. The cross tabulation for each variable was studied further to understand the nature of relationship between the biker and bike personality.

5.1.1 I Am Formal; My Bike Is Formal (Table 5.1)

As shown in Table 5.1, more than 90% of the total 3,106 respondents gave high ratings (4—agree or 5—strongly agree) to themselves and 88.1% of the

TABLE 5.1

Cross Tabulation of Biker–Bike Personality Ratings on the Personality Variable "Formal"

Case Processing Summary						
	Cases					
	Valid N	Percent	Missing	Percent	Total N	Percent
I am formal; my bike is formal	3106	100.0%	0	.0%	3106	100.0%

I am formal, my bike is formal—cross tabulation								
			My bike is formal					Total
			Strongly disagree	Disagree	Can't say	Agree	Strongly agree	
I am formal	Strongly disagree	Count	2	2	2	4	2	12
		% within I am formal	16.7%	16.7%	16.7%	33.3%	16.7%	100.0%
		% within My bike is formal	4.3%	1.0%	1.6%	.2%	.2%	.4%
		% of Total	0.1%	0.1%	0.1%	0.1%	0.1%	0.4%
	Disagree	Count	7	25	10	44	5	91
		% within I am formal	7.7%	27.5%	11.0%	48.4%	5.5%	100.0%
		% within My bike is formal	15.2%	12.9%	7.8%	2.3%	.6%	2.9%
		% of Total	0.2%	0.8%	0.3%	1.4%	0.2%	2.9%
	Can't Say	Count	6	20	33	117	24	200
		% within I am formal	3.0%	10.0%	16.5%	58.5%	12.0%	100.0%
		% within My bike is formal	13.0%	10.3%	25.8%	6.2%	2.8%	6.4%
		% of Total	0.2%	0.6%	1.1%	3.8%	0.8%	6.4%
	Agree	Count	20	108	70	1339	360	1897
		% within I am formal	1.1%	5.7%	3.7%	70.6%	19.0%	100.0%
		% within My bike is formal	43.5%	55.7%	54.7%	71.5%	41.7%	61.1%
		% of Total	0.6%	3.5%	2.3%	43.1%	11.6%	61.1%
	Strongly agree	Count	11	39	13	370	473	906
		% within I am formal	1.2%	4.3%	1.4%	40.8%	52.2%	100.0%
		% within My bike is formal	23.9%	20.1%	10.2%	19.7%	54.7%	29.2%
		% of Total	0.4%	1.3%	.4%	11.9%	15.2%	29.2%
Total		Count	46	194	128	1874	864	3106
		% within I am formal	1.5%	6.2%	4.1%	60.3%	27.8%	100.0%
		% within My bike is formal	100.0%	100.0%	100.0%	100.0%	100.0%	100.0%
		% of Total	1.5%	6.2%	4.1%	60.3%	27.8%	100.0%

Chi-square tests			
	Value	df	Asymp. Sig. (two-sided)
Pearson chi-square	608.706a	16	0
Likelihood ratio	506.447	16	0
Linear-by-linear association	304.56	1	0
N of valid cases	3106		
Expected count less than 5. The minimum expected count is 18.			

total respondents gave similar high ratings to their bikes on the personality variable "formal." This clearly shows that most respondents found themselves and their bikes formal in personality.

Study of the cross-tabulation table reveals that 81.2% of the respondents gave high ratings (4—agree or 5—strongly agree) to their own and their bike's personality. Only 1.8% of the total respondents gave lower ratings (1—strongly disagree or 2—disagree) to themselves and higher ratings (4—agree or 5—strongly agree) to their bikes on this personality variable. And only 5.8% respondents gave higher ratings (4—agree or 5—strongly agree) to themselves and lower ratings (1—strongly disagree or 2—disagree) to their bikes. Thus, only around 7.6% respondents gave clear incongruent ratings as compared to a majority of congruent ratings given by more than 88.1% (i.e., 2,542 out of 3,106) respondents. The findings show a clear congruence between the biker–bike personality ratings on this personality variable.

5.1.2 I Am Friendly; My Bike Is Friendly (Table 5.2)

As shown in Table 5.2, 97.5% of the total 3,106 respondents gave high ratings (4—agree or 5—strongly agree) to themselves and 96% of the total respondents gave similar ratings to their bikes on the personality variable "friendly." Only a marginal 0.7% and 1.8% of the total respondents gave low ratings (1—strongly disagree or 2—disagree) to themselves and their bikes, respectively. And 1.8% and 2.1% respondents chose the option "3—can't say" for rating their own and their bike's personality, respectively. This clearly shows that most respondents found themselves and their bikes friendly in personality.

Detailed analysis of the cross-tabulation table reveals that 84.1% of the respondents gave similar high ratings (4—agree and 5—strongly agree) to their own and their bike's personality. Only 0.6% of the total respondents gave lower ratings (1—strongly disagree or 2—disagree) to themselves and higher ratings (4—agree or 5—strongly agree) to their bikes on this personality variable. And only 1.6% respondents gave higher ratings (4—agree or 5—strongly agree) to themselves and lower ratings (1—strongly disagree or 2—disagree) to their bikes. Thus, only around 2.2% respondents gave clear incongruent ratings as compared to a majority of congruent ratings given by more than 84.1% (i.e., 2,612 out of 3,106) respondents. Similar to the findings for the previous personality variable (formal), clear congruence is seen between the biker–bike personality ratings for this personality variable as well.

5.1.3 I Am Comfortable; My Bike Is Comfortable (Table 5.3)

As shown in Table 5.3, 91% of the total 3,106 respondents gave high ratings (4—agree or 5—strongly agree) to themselves and even larger percentage (93.9%) of the total respondents gave similar ratings to their bikes on the

TABLE 5.2

Cross Tabulation of Biker–Bike Personality Ratings on the Personality Variable "Friendly"

Case Processing Summary						
	Cases					
	Valid N	Percent	Percent	Missing	Total N	Percent
I am friendly; my bike is friendly	3106	100.0%	0	.0%	3106	100.0%

I am friendly, my bike is friendly—cross tabulation								
			My bike is friendly					Total
			Strongly disagree	Disagree	Can't say	Agree	Strongly agree	
I am friendly	Strongly disagree	Count	0	0	0	0	3	3
		% within I am friendly	0.0%	0.0%	0.0%	0.0%	100.0%	100.0%
		% within My bike is friendly	0.0%	0.0%	0.0%	0.0%	0.2%	0.1%
		% of Total	0.0%	0.0%	0.0%	0.0%	0.1%	0.1%
	Disagree	Count	1	1	1	13	3	19
		% within I am friendly	5.3%	5.3%	5.3%	68.4%	15.8%	100.0%
		% within My bike is friendly	7.1%	2.6%	1.6%	0.8%	0.2%	0.6%
		% of Total	0.0%	0.0%	0.0%	0.4%	0.1%	0.6%
	Can't say	Count	1	0	5	34	15	55
		% within I am friendly	1.8%	0.0%	9.1%	61.8%	27.3%	100.0%
		% within My bike is friendly	7.1%	0.0%	7.8%	2.0%	1.2%	1.8%
		% of Total	0.0%	0.0%	0.2%	1.1%	0.5%	1.8%
	Agree	Count	6	26	30	1006	365	1433
		% within I am friendly	0.4%	1.8%	2.1%	70.2%	25.5%	100.0%
		% within My bike is friendly	42.9%	66.7%	46.9%	58.7%	28.6%	46.1%
		% of Total	0.2%	0.8%	1.0%	32.4%	11.8%	46.1%
	Strongly agree	Count	6	12	28	660	890	1596
		% within I am friendly	0.4%	0.8%	1.8%	41.4%	55.8%	100.0%
		% within My bike is friendly	42.9%	30.8%	43.8%	38.5%	69.7%	51.4%
		% of Total	0.2%	0.4%	0.9%	21.2%	28.7%	51.4%
Total		Count	14	39	64	1713	1276	3106
		% within I am friendly	0.5%	1.3%	2.1%	55.2%	41.1%	100.0%
		% within My bike is friendly	100.0%	100.0%	100.0%	100.0%	100.0%	100.0%
		% of Total	0.5%	1.3%	2.1%	55.2%	41.1%	100.0%

Chi-square tests			
	Value	df	Asymp. Sig. (2—sided)
Pearson chi-square	330.631a	16	0
Likelihood ratio	324.078	16	0
Linear-by-linear association	196.391	1	0
N of valid cases	3106		
Expected count less than 5. The minimum expected count is .01.			

TABLE 5.3

Cross Tabulation of Biker–Bike Personality Ratings on the Personality Variable "Comfortable"

Case Processing Summary						
	Cases					
	Valid N	Percent	Missing	Percent	Total N	Percent
I am comfortable; my bike is comfortable	3106	100.0%	0	0.0%	3106	100.0%

I am comfortable, my bike is comfortable—cross tabulation								
			My bike is comfortable					Total
			Strongly disagree	Disagree	Can't say	Agree	Strongly agree	
I am comfortable	Strongly disagree	Count	0	0	1	2	4	7
		% within I am comfortable	0.0%	0.0%	14.3%	28.6%	57.1%	100.0%
		% within My bike is comfortable	0.0%	0.0%	1.1%	0.1%	0.3%	0.2%
		% of Total	0.0%	0.0%	0.0%	0.1%	0.1%	0.2%
	Disagree	Count	0	3	2	37	32	74
		% within I am comfortable	0.0%	4.1%	2.7%	50.0%	43.2%	100.0%
		% within My bike is comfortable	0.0%	3.7%	2.2%	2.4%	2.4%	2.4%
		% of Total	0.0%	0.1%	0.1%	1.2%	1.0%	2.4%
	Can't say	Count	2	4	26	114	65	211
		% within I am comfortable	0.9%	1.9%	12.3%	54.0%	30.8%	100.0%
		% within My bike is comfortable	11.1%	4.9%	28.6%	7.3%	4.8%	6.8%
		% of Total	0.1%	0.1%	0.8%	3.7%	2.1%	6.8%
	Agree	Count	11	47	52	1075	620	1805
		% within I am comfortable	0.6%	2.6%	2.9%	59.6%	34.3%	100.0%
		% within My bike is comfortable	61.1%	57.3%	57.1%	69.0%	45.7%	58.1%
		% of Total	0.4%	1.5%	1.7%	34.6%	20.0%	58.1%
	Strongly agree	Count	5	28	10	331	635	1009
		% within I am comfortable	0.5%	2.8%	1.0%	32.8%	62.9%	100.0%
		% within My bike is comfortable	27.8%	34.1%	11.0%	21.2%	46.8%	32.5%
		% of Total	0.2%	0.9%	0.3%	10.7%	20.4%	32.5%
Total		Count	18	82	91	1559	1356	3106
		% within I am comfortable	0.6%	2.6%	2.9%	50.2%	43.7%	100.0%
		% within My bike is comfortable	100.0%	100.0%	100.0%	100.0%	100.0%	100.0%
		% of Total	0.6%	2.6%	2.9%	50.2%	43.7%	100.0%

Chi-square tests			
	Value	df	Asymp. Sig. (2—sided)
Pearson chi-square	305.526a	16	0
Likelihood ratio	280.809	16	0
Linear-by-linear association	91.261	1	0
N of valid cases	3106		
Expected count less than 5. The minimum expected count is .04.			

personality variable "comfortable." Only a marginal 2.6% and 3.2% of the total respondents gave low ratings (1—strongly disagree or 2—disagree) to themselves and their bikes, respectively. And 6.8% and 2.9% respondents chose the option "3—can't say" for rating their own and their bike's personality, respectively. This clearly shows that most respondents found themselves and their bikes comfortable in personality.

An exhaustive analysis of the cross-tabulation table reveals that 85.7% of the respondents gave similar high ratings (4—agree or 5—strongly agree) to their own and their bike's personality. Only 2.4% of the total respondents gave lower ratings (1—strongly disagree or 2—disagree) to themselves and higher ratings (4—agree or 5—strongly agree) to their bikes on this personality variable. And 3% respondents gave higher ratings (4—agree or 5—strongly agree) to themselves and lower ratings (1—strongly disagree or 2—disagree) to their bikes. Thus, only around 5.4% respondents gave clear incongruent ratings as compared to a majority of congruent ratings given by more than 85.7% (i.e., 2,661 out of 3,106) respondents. Similar to the findings for the previous personality variables, a clear congruence is seen between the biker–bike personality ratings for this personality variable as well.

5.1.4 I Am Professional; My Bike Is Professional (Table 5.4)

As shown in Table 5.4, 90% of the total 3,106 respondents gave high ratings (4—agree or 5—strongly agree) to themselves and 87.3% of the total respondents gave similar ratings to their bikes on the personality variable "professional." Only a marginal 2.5% and 5.8% of the total respondents gave low ratings (1—strongly disagree or 2—disagree) to themselves and their bikes, respectively. And 7.6% and 10.6% respondents chose the option "3—can't say" for rating their own and their bike's personality, respectively. This clearly shows that most respondents found themselves and their bikes as professional in personality.

Detailed analysis of the cross-tabulation table reveals that 76.7% of the respondents gave similar high ratings (4—agree or 5—strongly agree) to their own and their bike's personality. Only 1.6% of the total respondents gave lower ratings (1—strongly disagree or 2—disagree) to themselves and higher ratings (4—agree or 5—strongly agree) to their bikes on this personality variable. And 4.4% respondents gave higher ratings (4—agree or 5—strongly agree) to themselves and lower ratings (1—strongly disagree or 2—disagree) to their bikes. Thus, only around 6% respondents gave clear incongruent ratings as compared to a majority of congruent ratings given by more than 76.7% (i.e., 2,384 out of 3,106) respondents. A high rating for biker personality was expected on this personality variable as more than 80% of the sample population was employed in the public or private sector, and around 11% were engaged in business. However, similar high ratings for bike further highlight the congruent relationship between biker–bike personalities for this personality variable as well.

TABLE 5.4

Cross Tabulation of Biker–Bike Personality Ratings on the Personality Variable "Professional"

Case Processing Summary						
	Cases					
	Valid N	Percent	Missing	Percent	Total N	Percent
I am professional; my bike is professional	3106	100.0%	0	.0%	3106	100.0%

I am professional, my bike is professional—cross tabulation								
			My bike is professional					Total
			Strongly disagree	Disagree	Can't say	Agree	Strongly agree	
I am professional	Strongly disagree	Count	1	4	2	1	1	9
		% within I am professional	11.1%	44.4%	22.2%	11.1%	11.1%	100.0%
		% within My bike is professional	3.7%	2.6%	0.6%	0.1%	0.1%	0.3%
		% of Total	0.0%	0.1%	0.1%	0.0%	0.0%	0.3%
	Disagree	Count	1	12	7	39	8	67
		% within I am professional	1.5%	17.9%	10.4%	58.2%	11.9%	100.0%
		% within My bike is professional	3.7%	7.9%	2.1%	2.3%	.9%	2.2%
		% of Total	0.0%	0.4%	0.2%	1.3%	0.3%	2.2%
	Can't say	Count	3	20	47	117	49	236
		% within I am professional	1.3%	8.5%	19.9%	49.6%	20.8%	100.0%
		% within My bike is professional	11.1%	13.2%	14.3%	7.0%	5.3%	7.6%
		% of Total	0.1%	0.6%	1.5%	3.8%	1.6%	7.6%
	Agree	Count	13	66	173	955	342	1549
		% within I am professional	0.8%	4.3%	11.2%	61.7%	22.1%	100.0%
		% within My bike is professional	48.1%	43.7%	52.6%	57.0%	37.1%	49.9%
		% of Total	0.4%	2.1%	5.6%	30.7%	11.0%	49.9%
	Strongly agree	Count	9	49	100	564	523	1245
		% within I am professional	0.7%	3.9%	8.0%	45.3%	42.0%	100.0%
		% within My bike is professional	33.3%	32.5%	30.4%	33.7%	56.7%	40.1%
		% of Total	0.3%	1.6%	3.2%	18.2%	16.8%	40.1%
Total		Count	27	151	329	1676	923	3106
		% within I am professional	0.9%	4.9%	10.6%	54.0%	29.7%	100.0%
		% within My bike is professional	100.0%	100.0%	100.0%	100.0%	100.0%	100.0%
		% of Total	0.9%	4.9%	10.6%	54.0%	29.7%	100.0%

Chi-Square Tests			
	Value	df	Asymp. Sig. (2-sided)
Pearson chi-square	249.437a	16	0
Likelihood ratio	209.787	16	0
Linear-by-linear association	131.308	1	0
Number of valid cases	3106		
Expected count less than 5. The minimum expected count is .08.			

5.1.5 I Am Heavy; My Bike Is Heavy (Table 5.5)

As shown in Table 5.5, 25% of the total 3,106 respondents gave higher ratings (4—agree or 5—strongly agree) to themselves and 29.4% of the total respondents gave similar ratings to their bikes on the personality variable "heavy." On the other hand, 58.8% and 54% of the total respondents gave low ratings (1—strongly disagree or 2—disagree) to themselves and their bikes, respectively. And 16.2% and 16.6% respondents chose the option "3—can't say" for rating their own and their bike's personality, respectively. Unlike the previously analyzed personality variables, this personality variable showed different rating trends. A mix of higher and lower ratings was reported. Also, unlike the previous personality variables, a majority of the respondents gave lower ratings to themselves and their bikes on this personality variable.

Detailed analysis of the cross-tabulation table reveals that 17%, 37.6%, and 4.7% of the total respondents gave similar high (4—agree or 5—strongly agree), low (1—strongly disagree or 2—disagree), and average (3—can't) ratings to their own and their bike's personality, respectively. However, 13.2% of the total respondents gave lower ratings (1—strongly disagree or 2—disagree) to themselves and higher ratings (4—agree or 5—strongly agree) to their bikes on this personality variable. And 10.4% respondents gave higher ratings (4—agree or 5—strongly agree) to themselves and lower ratings (1—strongly disagree or 2—disagree) to their bikes. Thus, around 23.6% respondents gave clear incongruent ratings as compared to a majority of congruent ratings given by more than 59.3% (i.e., 1,841 out of 3,106) respondents. While there was diversity in the response associated to the personality variable "heavy," yet there was a congruence in the individual responses related to biker–bike personalities. However, higher incongruent ratings were also seen in this case as compared to previous personality variables.

5.1.6 I Am Traditional; My Bike Is Traditional (Table 5.6)

As shown in Table 5.6, 59.2% of the total 3,106 respondents gave high ratings (4—agree or 5—strongly agree) to themselves and 58.5% of the total respondents gave similar ratings to their bikes on the personality variable "traditional." On the other hand, 20.9% and 22.2% of the total respondents gave low ratings (1—strongly disagree or 2—disagree) to themselves and their bikes, respectively. And 19.9% and 19.4% respondents chose the option "3—can't say" for rating their own and their bike's personality, respectively.

In-depth analysis of the cross-tabulation table reveals that 40.6%, 8.6%, and 5.7% of the total respondents gave similar high (4—agree or 5—strongly agree), low (1—strongly disagree or 2—disagree), and average (3—can't say) ratings to their own and their bike's personality, respectively. However, 8.3% of the total respondents gave lower ratings (1—strongly disagree or 2—disagree) to themselves and higher ratings (4—agree or 5—strongly agree) to their bikes on this personality variable. And 9.2% respondents gave higher

TABLE 5.5

Cross Tabulation of Biker–Bike Personality Ratings on the Personality Variable "Heavy"

Case Processing Summary						
	Cases					
	Valid N	Percent	Missing	Percent	Total N	Percent
I am HEAVY; my bike is HEAVY	3106	100.0%	0	0.0%	3106	100.0%

I am heavy, my bike is heavy—cross tabulation								
			My bike is heavy					Total
			Strongly disagree	Disagree	Can't say	Agree	Strongly agree	
I am heavy	Strongly disagree	Count	78	226	55	68	29	456
		% within I am heavy	17.1%	49.6%	12.1%	14.9%	6.4%	100.0%
		% within My bike is heavy	36.1%	15.5%	10.6%	9.7%	13.9%	14.7%
		% of Total	2.5%	7.3%	1.8%	2.2%	.9%	14.7%
	Disagree	Count	86	776	197	263	49	1371
		% within I am heavy	6.3%	56.6%	14.4%	19.2%	3.6%	100.0%
		% within My bike is heavy	39.8%	53.2%	38.1%	37.4%	23.4%	44.1%
		% of Total	2.8%	25.0%	6.3%	8.5%	1.6%	44.1%
	Can't say	Count	15	179	145	129	35	503
		% within I am heavy	3.0%	35.6%	28.8%	25.6%	7.0%	100.0%
		% within My bike is heavy	6.9%	12.3%	28.0%	18.3%	16.7%	16.2%
		% of Total	0.5%	5.8%	4.7%	4.2%	1.1%	16.2%
	Agree	Count	27	235	100	207	68	637
		% within I am heavy	4.2%	36.9%	15.7%	32.5%	10.7%	100.0%
		% within My bike is heavy	12.5%	16.1%	19.3%	29.4%	32.5%	20.5%
		% of Total	0.9%	7.6%	3.2%	6.7%	2.2%	20.5%
	Strongly agree	Count	10	44	20	37	28	139
		% within I am heavy	7.2%	31.7%	14.4%	26.6%	20.1%	100.0%
		% within My bike is heavy	4.6%	3.0%	3.9%	5.3%	13.4%	4.5%
		% of Total	0.3%	1.4%	0.6%	1.2%	0.9%	4.5%
Total		Count	216	1460	517	704	209	3106
		% within I am heavy	7.0%	47.0%	16.6%	22.7%	6.7%	100.0%
		% within My bike is heavy	100.0%	100.0%	100.0%	100.0%	100.0%	100.0%
		% of Total	7.0%	47.0%	16.6%	22.7%	6.7%	100.0%

Chi-square tests			
	Value	df	A Symp. Sig. (two-sided)
Pearson chi-square	326.414[a]	16	0
Likelihood ratio	292.945	16	0
Linear-by-linear association	169.747	1	0
Number of valid cases	3106		

Expected count less than 5. The minimum expected count is 9.35.

TABLE 5.6

Cross Tabulation of Biker–Bike Personality Ratings on the Personality Variable "Traditional"

Case Processing Summary						
	Cases					
	Valid N	Percent	Missing	Percent	Total N	Percent
I am traditional; my bike is traditional	3106	100.0%	0	0.0%	3106	100.0%

I am traditional, my bike is traditional—cross tabulation								
			My bike is traditional					Total
			Strongly disagree	Disagree	Can't say	Agree	Strongly agree	
I am traditional	Strongly disagree	Count	15	33	19	27	5	99
		% within I am traditional	15.2%	33.3%	19.2%	27.3%	5.1%	100.0%
		% within My bike is traditional	17.2%	5.5%	3.1%	1.9%	1.4%	3.2%
		% of Total	0.5%	1.1%	0.6%	0.9%	0.2%	3.2%
	Disagree	Count	25	192	111	187	36	551
		% within I am traditional	4.5%	34.8%	20.1%	33.9%	6.5%	100.0%
		% within My bike is traditional	28.7%	31.8%	18.1%	13.0%	10.0%	17.7%
		% of Total	0.8%	6.2%	3.6%	6.0%	1.2%	17.7%
	Can't say	Count	15	140	176	243	43	617
		% within I am traditional	2.4%	22.7%	28.5%	39.4%	7.0%	100.0%
		% within My bike is traditional	17.2%	23.2%	28.8%	16.8%	11.9%	19.9%
		% of Total	0.5%	4.5%	5.7%	7.8%	1.4%	19.9%
	Agree	Count	24	201	262	811	155	1453
		% within I am traditional	1.7%	13.8%	18.0%	55.8%	10.7%	100.0%
		% within My bike is traditional	27.6%	33.3%	42.8%	56.2%	42.9%	46.8%
		% of Total	0.8%	6.5%	8.4%	26.1%	5.0%	46.8%
	Strongly agree	Count	8	37	44	175	122	386
		% within I am traditional	2.1%	9.6%	11.4%	45.3%	31.6%	100.0%
		% within My bike is traditional	9.2%	6.1%	7.2%	12.1%	33.8%	12.4%
		% of Total	0.3%	1.2%	1.4%	5.6%	3.9%	12.4%
Total		Count	87	603	612	1443	361	3106
		% within I am traditional	2.8%	19.4%	19.7%	46.5%	11.6%	100.0%
		% within My bike is traditional	100.0%	100.0%	100.0%	100.0%	100.0%	100.0%
		% of Total	2.8%	19.4%	19.7%	46.5%	11.6%	100.0%

Chi-square tests			
	Value	df	Asymp. Sig. (two-sided)
Pearson chi-square	452.915a	16	0
Likelihood ratio	381.654	16	0
Linear-by-linear association	285.462	1	0
N of valid cases	3106		

Expected count less than 5. The minimum expected count is 2.77.

ratings (4—agree or 5—strongly agree) to themselves and lower ratings (1—strongly disagree or 2—disagree) to their bikes. Thus, around 17.5% respondents gave clear incongruent ratings as compared to a majority of congruent ratings given by more than 54.9% (i.e., 1,577 out of 3,106) respondents. While there was diversity in the response associated with the personality variable "traditional," yet there was a congruence in the individual responses related to biker–bike personalities.

5.1.7 I Am Popular; My Bike Is Popular (Table 5.7)

As shown in Table 5.7, 58.3% of the total 3,106 respondents gave high ratings (4—agree or 5—strongly agree) to themselves and 87% of the total respondents gave similar ratings to their bikes on the personality variable "popular." On the other hand, 7.9% and 4.9% of the total respondents gave low ratings (1—strongly disagree or 2—disagree) to themselves and their bikes, respectively. And 33.8% and 8.1% respondents chose the option "3—can't say" for rating their own and their bike's personality, respectively. This clearly shows that most respondents found themselves and their bikes to be popular in personality.

Detailed analysis of the cross-tabulation table reveals that 52%, 0.5%, and 3.6% of the total respondents gave similar high (4—agree or 5—strongly agree), low (1—strongly disagree or 2—disagree), and average (3—can't say) ratings to their own and their bike's personality, respectively. However, 6.7% of the total respondents gave lower ratings (1—strongly disagree or 2—disagree) to themselves and higher ratings (4—agree or 5—strongly agree) to their bikes on this personality variable. And 2.7% respondents gave higher ratings (4—agree or 5—strongly agree) to themselves and lower ratings (1—strongly disagree or 2—disagree) to their bikes. Thus, only around 9.4% respondents gave clear incongruent ratings as compared to a majority of congruent ratings given by more than 56.1% (i.e., 1,742 out of 3,106) respondents. Also a large percentage (28.4%) of respondents rated their bike as popular but chose "3—can't say" for rating their own popularity. All those responses where respondent chose 3—can't say option for rating either own or their bike's personality could not be clearly categorized as congruent or incongruent. Except for such 34.5% responses, the findings showed a clear congruence between the biker–bike personality ratings for this personality variable as well.

5.1.8 I Am Emotional; My Bike Is Emotional (Table 5.8)

As shown in Table 5.8, 71.3% of the total 3,106 respondents gave high ratings (4—agree or 5—strongly agree) to themselves and 39.3% of the total respondents gave similar ratings to their bikes on the personality variable "emotional." On the other hand, 13% and 21.8% of the total respondents gave low ratings (1—strongly disagree or 2—disagree) to themselves and their bikes, respectively. And 15.8% and 39% respondents chose the option "3—can't

TABLE 5.7

Cross Tabulation of Biker–Bike Personality Ratings on the Personality Variable "Popular"

Case Processing Summary						
	Cases					
	Valid N	Percent	Missing	Percent	Total N	Percent
I am popular; my bike is popular	3106	100.0%	0	0.0%	3106	100.0%

I am popular; my bike is popular—cross tabulation								
			My bike is popular					Total
			Strongly disagree	Disagree	Can't say	Agree	Strongly agree	
I am popular	Strongly disagree	Count	0	0	2	14	5	21
		% within I am popular	0.0%	0.0%	9.5%	66.7%	23.8%	100.0%
		% within My bike is popular	0.0%	0.0%	0.8%	1.0%	0.4%	0.7%
		% of Total	0.0%	0.0%	0.1%	0.5%	0.2%	0.7%
	Disagree	Count	1	14	23	109	78	225
		% within I am popular	0.4%	6.2%	10.2%	48.4%	34.7%	100.0%
		% within My bike is popular	4.8%	10.6%	9.1%	8.1%	5.8%	7.2%
		% of Total	0.0%	0.5%	0.7%	3.5%	2.5%	7.2%
	Can't say	Count	6	50	111	474	408	1049
		% within I am popular	0.6%	4.8%	10.6%	45.2%	38.9%	100.0%
		% within My bike is popular	28.6%	37.9%	44.0%	35.1%	30.2%	33.8%
		% of Total	0.2%	1.6%	3.6%	15.3%	13.1%	33.8%
	Agree	Count	8	54	91	633	593	1379
		% within I am popular	0.6%	3.9%	6.6%	45.9%	43.0%	100.0%
		% within My bike is popular	38.1%	40.9%	36.1%	46.9%	43.9%	44.4%
		% of Total	0.3%	1.7%	2.9%	20.4%	19.1%	44.4%
	Strongly agree	Count	6	14	25	120	267	432
		% within I am popular	1.4%	3.2%	5.8%	27.8%	61.8%	100.0%
		% within My bike is popular	28.6%	10.6%	9.9%	8.9%	19.8%	13.9%
		% of Total	0.2%	0.5%	0.8%	3.9%	8.6%	13.9%
Total		Count	21	132	252	1350	1351	3106
		% within I am popular	0.7%	4.2%	8.1%	43.5%	43.5%	100.0%
		% within My bike is popular	100.0%	100.0%	100.0%	100.0%	100.0%	100.0%
		% of Total	0.7%	4.2%	8.1%	43.5%	43.5%	100.0%

Chi-square tests			
	Value	df	Asymp. sig. (two-sided)
Pearson chi-square	100.199a	16	0
Likelihood ratio	100.278	16	0
Linear-by-linear association	39.208	1	0
Number of valid cases	3106		
Expected count less than 5. The minimum expected count is 0.14.			

TABLE 5.8

Cross Tabulation of Biker–Bike Personality Ratings on the Personality Variable "Emotional"

Case Processing Summary						
	Cases					
	Valid N	Percent	Missing	Percent	Total N	Percent
I am emotional; my bike is emotional	3106	100.0%	0	.0%	3106	100.0%

I am emotional, my bike is emotional—cross tabulation								
			My bike is total emotional					Total
			Strongly disagree	Disagree	Can't say	Agree	Strongly agree	
I am emotional	Strongly disagree	Count	5	13	18	9	4	49
		% within I am emotional	10.2%	26.5%	36.7%	18.4%	8.2%	100.0%
		% within My bike is emotional	5.4%	2.2%	1.5%	0.9%	1.5%	1.6%
		% of Total	0.2%	0.4%	0.6%	0.3%	0.1%	1.6%
	Disagree	Count	19	132	123	65	14	353
		% within I am emotional	5.4%	37.4%	34.8%	18.4%	4.0%	100.0%
		% within My bike is emotional	20.7%	22.6%	10.2%	6.8%	5.3%	11.4%
		% of Total	0.6%	4.2%	4.0%	2.1%	0.5%	11.4%
	Can't say	Count	16	91	249	105	29	490
		% within I am emotional	3.3%	18.6%	50.8%	21.4%	5.9%	100.0%
		% within My bike is emotional	17.4%	15.6%	20.6%	11.0%	10.9%	15.8%
		% of Total	0.5%	2.9%	8.0%	3.4%	0.9%	15.8%
	Agree	Count	37	261	601	561	100	1560
		% within I am emotional	2.4%	16.7%	38.5%	36.0%	6.4%	100.0%
		% within My bike is emotional	40.2%	44.7%	49.7%	58.8%	37.6%	50.2%
		% of Total	1.2%	8.4%	19.3%	18.1%	3.2%	50.2%
	Strongly agree	Count	15	87	219	214	119	654
		% within I am emotional	2.3%	13.3%	33.5%	32.7%	18.2%	100.0%
		% within My bike is emotional	16.3%	14.9%	18.1%	22.4%	44.7%	21.1%
		% of Total	0.5%	2.8%	7.1%	6.9%	3.8%	21.1%
Total		Count	92	584	1210	954	266	3106
		% within I am emotional	3.0%	18.8%	39.0%	30.7%	8.6%	100.0%
		% within My bike is emotional	100.0%	100.0%	100.0%	100.0%	100.0%	100.0%
		% of Total	3.0%	18.8%	39.0%	30.7%	8.6%	100.0%

Chi-square tests			
	Value	df	Asymp. Sig. (two-sided)
Pearson chi-square	264.000a	16	0
Likelihood ratio	234.99	16	0
Linear-by-linear association	152.705	1	0
N of valid cases	3106		

Expected count less than 5. The minimum expected count is 1.45.

say" for rating their own and their bike's personality, respectively. A clear majority of respondents found themselves to be "emotional." However, the respondents were less sure when it came to rating their bikes on this personality variable. Though a large percentage (39.3%) rated their bikes as emotional, an equal percentage of respondents chose the option 3—can't say while rating their bike's personality.

An exhaustive analysis of the cross-tabulation table reveals that 32%, 5.4%, and 8% of the total respondents gave similar high (4—agree or 5—strongly agree), low (1—strongly disagree or 2—disagree), and average (3—can't say) ratings to their own and their bike's personality, respectively. However, 3% of the total respondents gave lower ratings (1—strongly disagree or 2—disagree) to themselves and higher ratings (4—agree or 5—strongly agree) to their bikes on this personality variable. And 12.9% respondents gave higher ratings (4—agree or 5—strongly agree) to themselves and lower ratings (1—strongly disagree or 2—disagree) to their bikes. Thus, around 15.9% respondents gave clear incongruent ratings as compared to a majority of congruent ratings given by more than 45.4% (i.e., 1,410 out of 3,106) respondents. Also, a large percentage (26.4%) of respondents rated themselves as emotional but chose 3—can't say for rating their bike as "emotional." All those responses where respondent chose "3—can't say" option for rating either own or bike personality could not be clearly categorized as congruent or incongruent. Except for such 38.7% responses, the findings showed a clear congruence between the biker–bike personality ratings for this personality variable as well.

5.1.9 I Am Mean; My Bike Is Mean (Table 5.9)

As shown in Table 5.9, 35.2% of the total 3,106 respondents gave high rating (4—agree or 5—strongly agree) to themselves and 32.5% of the total respondents gave similar ratings to their bikes on the personality variable "mean." On the other hand, 42.5% and 32.1% of the total respondents gave low ratings (1—strongly disagree or 2—disagree) to themselves and their bikes, respectively. And 22.3% and 35.5% respondents chose the option "3—can't say" for rating their own and their bike's personality, respectively. A diverse trend is seen for ratings given for biker–bike personality on this personality variable.

Analysis of the cross-tabulation table reveals that 18.8%, 23.6%, and 11.5% of the total respondents gave similar high (4—agree or 5—strongly agree), low (1—strongly disagree or 2—disagree), and average (3—can't say) ratings to their own and their bike's personality, respectively. However, 7% of the total respondents gave lower ratings (1—strongly disagree or 2—disagree) to themselves and higher ratings (4—agree or 5—strongly agree) to their bikes on this personality variable. And 4.2% respondents gave higher ratings (4—agree or 5—strongly agree) to themselves and lower ratings (1—strongly disagree or 2—disagree) to their bikes. Thus, around 11.2% respondents gave clear incongruent ratings as compared to a majority of congruent ratings

TABLE 5.9

Cross Tabulation of Biker–Bike Personality Ratings on the Personality Variable "Mean"

Case Processing Summary						
	Cases					
	Valid N	Percent	Missing	Percent	Total N	Percent
I am mean; my bike is mean	3106	100.0%	0	.0%	3106	100.0%

I am mean, my bike is mean—cross tabulation								
			My bike is mean					Total
			Strongly disagree	Disagree	Can't say	Agree	Strongly agree	
I am mean	Strongly disagree	Count	114	183	83	32	20	432
		% within I am mean	26.4%	42.4%	19.2%	7.4%	4.6%	100.0%
		% within My bike is mean	55.9%	23.1%	7.5%	4.1%	8.5%	13.9%
		% of Total	3.7%	5.9%	2.7%	1.0%	.6%	13.9%
	Disagree	Count	47	389	284	129	38	887
		% within I am mean	5.3%	43.9%	32.0%	14.5%	4.3%	100.0%
		% within My bike is mean	23.0%	49.1%	25.8%	16.7%	16.2%	28.6%
		% of Total	1.5%	12.5%	9.1%	4.2%	1.2%	28.6%
	Can't say	Count	21	111	358	154	50	694
		% within I am mean	3.0%	16.0%	51.6%	22.2%	7.2%	100.0%
		% within My bike is mean	10.3%	14.0%	32.5%	19.9%	21.3%	22.3%
		% of Total	0.7%	3.6%	11.5%	5.0%	1.6%	22.3%
	Agree	Count	13	91	307	382	80	873
		% within I am mean	1.5%	10.4%	35.2%	43.8%	9.2%	100.0%
		% within My bike is mean	6.4%	11.5%	27.9%	49.5%	34.0%	28.1%
		% of Total	0.4%	2.9%	9.9%	12.3%	2.6%	28.1%
	Strongly agree	Count	9	19	70	75	47	220
		% within I am mean	4.1%	8.6%	31.8%	34.1%	21.4%	100.0%
		% within My bike is mean	4.4%	2.4%	6.4%	9.7%	20.0%	7.1%
		% of Total	0.3%	0.6%	2.3%	2.4%	1.5%	7.1%
Total		Count	204	793	1102	772	235	3106
		% within I am mean	6.6%	25.5%	35.5%	24.9%	7.6%	100.0%
		% within My bike is mean	100.0%	100.0%	100.0%	100.0%	100.0%	100.0%
		% of Total	6.6%	25.5%	35.5%	24.9%	7.6%	100.0%

Chi-square tests			
	Value	df	Asymp. Sig. (two-sided)
Pearson chi-square	990.784a	16	0
Likelihood ratio	891.853	16	0
Linear-by-linear association	591.759	1	0
Number of valid cases	3106		

Expected count less than 5. The minimum expected count is 14.45.

given by more than 53.9% (i.e., 1,674 out of 3,106) respondents. Also, a large percentage (24%) of respondents chose the "3—can't say" option for rating their bikes while attributing higher (4—agree or 5—strongly agree) or lower (1—strongly disagree or 2—disagree) ratings to themselves. All those responses where respondents chose "3—can't say" option for rating either own or bike's personality could not be clearly categorized as congruent or incongruent. Except for such 34.9% responses, the findings show a clear congruence between the biker–bike personality ratings for this personality variable as well.

5.1.10 I Am Aggressive; My Bike Is Aggressive (Table 5.10)

As shown in Table 5.10, 47.3% of the total 3,106 respondents gave high ratings (4—agree or 5—strongly agree) to themselves, and 57.2% of the total respondents gave similar ratings to their bikes on the personality variable "aggressive." On the other hand, 31.2% and 22.8% of the total respondents gave low ratings (1—strongly disagree or 2—disagree) to themselves and their bikes, respectively. And 21.5% and 19.9% respondents chose the option "3—can't say" for rating their own and their bike's personality, respectively. A diverse trend is seen for ratings given for biker–bike personality on this personality variable.

A study of the cross-tabulation table reveals that 32.2%, 11.3%, and 6.1% of the total respondents gave similar high (4—agree or 5—strongly agree), low (1—strongly disagree or 2—disagree), and average (3—can't say) ratings to their own and their bike's personality, respectively. However, 13.7% of the total respondents gave lower ratings (1—strongly disagree or 2—disagree) to themselves and higher ratings (4—agree or 5—strongly agree) to their bikes on this personality attribute. And 7.6% respondents gave higher ratings (4—agree or 5—strongly agree) to themselves and lower ratings (1—strongly disagree or 2—disagree) to their bikes. Thus, around 21.3% respondents gave clear incongruent ratings as compared to a majority of congruent ratings given by more than 49.6% (i.e., 1,540 out of 3,106) respondents. Also, a large percentage (15.3%) of respondents chose the "3—can't say" option for rating their bikes while attributing higher (4—agree or 5—strongly agree) or lower (1—strongly disagree or 2—disagree) ratings to themselves. Similarly, 13.7% of respondents chose the "3—can't say" option for rating themselves while attributing higher (4—agree or 5—strongly agree) or lower (1—strongly disagree or 2—disagree) ratings to their bikes. Except for these 29% (15.3% + 13.7% = 29%) responses, which couldn't be clearly categorized as congruent or incongruent, the findings showed congruence between the biker–bike personality ratings for 49.6% of the responses. However, just like the personality variable "heavy," a large percentage of incongruent (21.3%) ratings were reported by participants of the survey on this personality variable as well.

TABLE 5.10

Cross Tabulation of Biker–Bike Personality Ratings on the Personality Variable "Aggressive"

Case Processing Summary						
	Cases					
	Valid N	Percent	Missing	Percent	Total N	Percent
I am aggressive; my bike is aggressive	3106	100.0%	0	.0%	3106	100.0%

I am aggressive, my bike is aggressive—cross tabulation								
			My bike is aggressive					Total
			Strongly disagree	Disagree	Can't say	Agree	Strongly agree	
I am aggressive	Strongly disagree	Count	22	59	28	56	31	196
		% within I am aggressive	11.2%	30.1%	14.3%	28.6%	15.8%	100.0%
		% within My bike is aggressive	22.7%	9.6%	4.5%	4.5%	5.9%	6.3%
		% of Total	0.7%	1.9%	.9%	1.8%	1.0%	6.3%
	Disagree	Count	33	235	166	264	75	773
		% within I am aggressive	4.3%	30.4%	21.5%	34.2%	9.7%	100.0%
		% within My bike is aggressive	34.0%	38.3%	26.9%	21.0%	14.3%	24.9%
		% of Total	1.1%	7.6%	5.3%	8.5%	2.4%	24.9%
	Can't say	Count	13	113	191	264	87	668
		% within I am aggressive	1.9%	16.9%	28.6%	39.5%	13.0%	100.0%
		% within My bike is aggressive	13.4%	18.4%	30.9%	21.0%	16.6%	21.5%
		% of Total	0.4%	3.6%	6.1%	8.5%	2.8%	21.5%
	Agree	Count	19	167	186	545	200	1117
		% within I am aggressive	1.7%	15.0%	16.7%	48.8%	17.9%	100.0%
		% within My bike is aggressive	19.6%	27.2%	30.1%	43.4%	38.2%	36.0%
		% of Total	0.6%	5.4%	6.0%	17.5%	6.4%	36.0%
	Strongly agree	Count	10	39	47	126	130	352
		% within I am aggressive	2.8%	11.1%	13.4%	35.8%	36.9%	100.0%
		% within My bike is aggressive	10.3%	6.4%	7.6%	10.0%	24.9%	11.3%
		% of Total	0.3%	1.3%	1.5%	4.1%	4.2%	11.3%
Total		Count	97	613	618	1255	523	3106
		% within I am aggressive	3.1%	19.7%	19.9%	40.4%	16.8%	100.0%
		% within My bike is aggressive	100.0%	100.0%	100.0%	100.0%	100.0%	100.0%
		% of Total	3.1%	19.7%	19.9%	40.4%	16.8%	100.0%

Chi-square tests			
	Value	df	Asymp. Sig. (two-sided)
Pearson chi-square	332.033a	16	0
Likelihood ratio	294.708	16	0
Linear—by—linear association	189.847	1	0
Number of valid cases	3106		
Expected count less than 5. The minimum expected count is 6.12.			

5.1.11 I Am Sporty; My Bike Is Sporty (Table 5.11)

As shown in Table 5.11, 70.6% of the total 3,106 respondents gave high ratings (4—agree or 5—strongly agree) to themselves, and 67.6% of the total respondents gave similar ratings to their bikes on the personality variable "sporty." On the other hand, 13.6% and 18.9% of the total respondents gave low ratings (1—strongly disagree or 2—disagree) to themselves and their bikes, respectively. And 15.8% and 13.4% respondents chose the option "3—can't say" for rating their own and their bike's personality, respectively. This clearly shows that most respondents found themselves and their bikes to be sporty in personality.

Detailed analysis of the cross-tabulation table reveals that 52.7%, 4.7%, and 3.8% of the total respondents gave similar high (4—agree or 5—strongly agree), low (1—strongly disagree or 2—disagree), and average (3—can't say) ratings to their own and their bike's personality, respectively. However, 6.6% of the total respondents gave lower ratings (1—strongly disagree or 2—disagree) to themselves and higher ratings (4—agree or 5—strongly agree) to their bikes on this personality variable. And 10.6% respondents gave higher ratings (4—agree or 5—strongly agree) to themselves and lower ratings (1—strongly disagree or 2—disagree) to their bikes. Thus, around 17.2% respondents gave clear incongruent ratings as compared to a majority of congruent ratings given by more than 61.2% (i.e., 1,900 out of 3,106) respondents. Thus, a clear congruence is seen between the biker–bike personality ratings for this personality variable.

5.1.12 I Am Sharp; My Bike Is Sharp (Table 5.12)

As shown in Table 5.12, 77.9% of the total 3,106 respondents gave high ratings (4—agree or 5—strongly agree) to themselves and 67.4% of the total respondents gave similar ratings to their bikes on the personality variable "sharp." On the other hand, 3.9% and 13.3% of the total respondents gave low ratings (1—strongly disagree or 2—disagree) to themselves and their bikes, respectively. And 18.2% and 19.3% respondents chose the option "3—can't say" for rating their own and their bike's personality, respectively. This clearly shows that most respondents found themselves and their bikes to be sharp in personality.

Detailed analysis of the cross-tabulation table reveals that 55.6%, 0.9%, and 5.9% of the total respondents gave similar high (4—agree or 5—strongly agree), low (1—strongly disagree or 2—disagree), and average (3—can't say) ratings to their own and their bike's personality, respectively. However, 2.1% of the total respondents gave lower ratings (1—strongly disagree or 2—disagree) to themselves and higher ratings (4—agree or 5—strongly agree) to their bikes on this personality variable. And 9.9% of the respondents gave higher ratings (4—agree or 5—strongly agree) to themselves and lower ratings (1—strongly disagree or 2—disagree) to their bikes. Thus, around 12%

TABLE 5.11

Cross Tabulation of Biker–Bike Personality Ratings on the Personality Variable "Sporty"

Case Processing Summary						
	Cases					
	Valid N	Percent	Missing	Percent	Total N	Percent
I am sporty; my bike is sporty	3106	100.0%	0	.0%	3106	100.0%

I am sporty, my bike is sporty—cross tabulation								
			My bike is sporty					Total
			Strongly disagree	Disagree	Can't say	Agree	Strongly agree	
I am sporty	Strongly disagree	Count	5	11	4	15	8	43
		% within I am sporty	11.6%	25.6%	9.3%	34.9%	18.6%	100.0%
		% within My bike is sporty	6.7%	2.1%	1.0%	1.0%	1.3%	1.4%
		% of Total	0.2%	0.4%	0.1%	0.5%	0.3%	1.4%
	Disagree	Count	17	112	69	138	42	378
		% within I am sporty	4.5%	29.6%	18.3%	36.5%	11.1%	100.0%
		% within My bike is sporty	22.7%	21.9%	16.5%	9.4%	6.6%	12.2%
		% of Total	0.5%	3.6%	2.2%	4.4%	1.4%	12.2%
	Can't say	Count	9	101	117	207	58	492
		% within I am sporty	1.8%	20.5%	23.8%	42.1%	11.8%	100.0%
		% within My bike is sporty	12.0%	19.7%	28.1%	14.1%	9.1%	15.8%
		% of Total	0.3%	3.3%	3.8%	6.7%	1.9%	15.8%
	Agree	Count	32	231	190	891	273	1617
		% within I am sporty	2.0%	14.3%	11.8%	55.1%	16.9%	100.0%
		% within My bike is sporty	42.7%	45.1%	45.6%	60.9%	42.8%	52.1%
		% of Total	1.0%	7.4%	6.1%	28.7%	8.8%	52.1%
	Strongly agree	Count	12	57	37	213	257	576
		% within I am sporty	2.1%	9.9%	6.4%	37.0%	44.6%	100.0%
		% within My bike is sporty	16.0%	11.1%	8.9%	14.5%	40.3%	18.5%
		% of Total	0.4%	1.8%	1.2%	6.9%	8.3%	18.5%
Total		Count	75	512	417	1464	638	3106
		% within I am sporty	2.4%	16.5%	13.4%	47.1%	20.5%	100.0%
		% within My bike is sporty	100.0%	100.0%	100.0%	100.0%	100.0%	100.0%
		% of Total	2.4%	16.5%	13.4%	47.1%	20.5%	100.0%

Chi-square tests			
	Value	df	Asymp. Sig. (two-sided)
Pearson chi-square	416.892a	16	0
Likelihood ratio	367.06	16	0
Linear-by-linear association	207.26	1	0
Number of valid cases	3106		
Expected count less than 5. The minimum expected count is 1.04.			

TABLE 5.12

Cross Tabulation of Biker–Bike Personality Ratings on the Personality Variable "Sharp"

Case Processing Summary						
	Cases					
	Valid N	Percent	Missing	Percent	Total N	Percent
I am sharp; my bike is sharp	3106	100.0%	0	0.0%	3106	100.0%

I am sharp, my bike is sharp—cross tabulation								
			My bike is sharp					Total
			Strongly disagree	Disagree	Can't say	Agree	Strongly agree	
I am sharp	Strongly disagree	Count	1	0	0	4	2	7
		% within I am sharp	14.3%	.0%	.0%	57.1%	28.6%	100.0%
		% within My bike is sharp	2.3%	0.0%	0.0%	0.3%	0.3%	0.2%
		% of Total	0.0%	0.0%	0.0%	0.1%	0.1%	0.2%
	Disagree	Count	2	25	30	50	8	115
		% within I am sharp	1.7%	21.7%	26.1%	43.5%	7.0%	100.0%
		% within My bike is sharp	4.7%	6.8%	5.0%	3.3%	1.4%	3.7%
		% of Total	0.1%	0.8%	1.0%	1.6%	0.3%	3.7%
	Can't say	Count	5	74	182	245	59	565
		% within I am sharp	0.9%	13.1%	32.2%	43.4%	10.4%	100.0%
		% within My bike is sharp	11.6%	20.1%	30.4%	16.1%	10.2%	18.2%
		% of Total	0.2%	2.4%	5.9%	7.9%	1.9%	18.2%
	Agree	Count	24	217	331	978	274	1824
		% within I am sharp	1.3%	11.9%	18.1%	53.6%	15.0%	100.0%
		% within My bike is sharp	55.8%	58.8%	55.3%	64.4%	47.6%	58.7%
		% of Total	0.8%	7.0%	10.7%	31.5%	8.8%	58.7%
	Strongly agree	Count	11	53	56	242	233	595
		% within I am sharp	1.8%	8.9%	9.4%	40.7%	39.2%	100.0%
		% within My bike is sharp	25.6%	14.4%	9.3%	15.9%	40.5%	19.2%
		% of Total	0.4%	1.7%	1.8%	7.8%	7.5%	19.2%
Total		Count	43	369	599	1519	576	3106
		% within I am sharp	1.4%	11.9%	19.3%	48.9%	18.5%	100.0%
		% within My bike is sharp	100.0%	100.0%	100.0%	100.0%	100.0%	100.0%
		% of Total	1.4%	11.9%	19.3%	48.9%	18.5%	100.0%

Chi-square tests			
	Value	df	Asymp. Sig. (two-sided)
Pearson chi-square	308.403a	16	0
Likelihood ratio	276.424	16	0
Linear-by-linear association	117.861	1	0
Number of valid cases	3106		
Expected count less than 5. The minimum expected count is 0.10.			

of the respondents gave clear incongruent ratings as compared to a majority of congruent ratings given by more than 62.4% (i.e., 1,938 out of 3,106) of the respondents. Thus, a clear congruence is seen between the biker–bike personality ratings for this personality variable.

5.1.13 I Am International; My Bike Is International (Table 5.13)

As shown in Table 5.13, 40.2% of the total 3,106 respondents gave high ratings (4—agree or 5—strongly agree) to themselves, and 56.1% of the total respondents gave similar ratings to their bikes on the personality variable "international." On the other hand, 28% and 15.2% of the total respondents gave low ratings (1—strongly disagree or 2—disagree) to themselves and their bikes, respectively. And 31.7% and 28.8% of the respondents chose the option "3—can't say" for rating their own and their bike's personality, respectively. A diverse trend is seen for ratings given for biker–bike personality on this personality variable.

The analysis of the cross tabulation table reveals that 27%, 6%, and 11.8% of the total respondents gave similar high (4—agree or 5—strongly agree), low (1—strongly disagree or 2—disagree), and average (3—can't say) ratings to their own and their bike's personality, respectively. However, 13.1% of the total respondents gave lower ratings (1—strongly disagree or 2—disagree) to themselves and higher ratings (4—agree or 5—strongly agree) to their bikes on this personality variable. And 5.1% of the respondents gave higher ratings (4—agree or 5—strongly agree) to themselves and lower ratings (1—strongly disagree or 2—disagree) to their bikes. Thus, around 18.2% of the respondents gave clear incongruent ratings as compared to a majority of congruent ratings given by more than 44.8% (i.e., 1,391 out of 3,106) of the respondents. Also, a large percentage (17%) of respondents chose the "3—can't say" option for rating their bikes while attributing higher (4—agree or 5—strongly agree) or lower (1—strongly disagree or 2—disagree) ratings to themselves. Similarly, 19.9% of the respondents chose the "3—can't say" option for rating themselves while attributing higher (4—agree or 5—strongly agree) or lower (1—strongly disagree or 2—disagree) ratings to their bikes. Except for these 37% (17% + 19.9%) responses, that could not be clearly categorized as congruent or incongruent, the findings showed congruence between the biker–bike personality ratings for 44.8% of the responses.

5.1.14 I Am Extraordinary; My Bike Is Extraordinary (Table 5.14)

As shown in Table 5.14, 35.8% of the total 3,106 respondents gave high ratings (4—agree or 5—strongly agree) to themselves, and 50.1% of the total respondents gave similar ratings to their bikes on the personality variable "extraordinary." On the other hand, 25.2% and 23.3% of the total respondents gave low ratings (1—strongly disagree or 2—disagree) to themselves and their bikes, respectively. And 39% and 26.6% respondents chose the option

TABLE 5.13

Cross Tabulation of Biker–Bike Personality Ratings on the Personality Variable "International"

Case Processing Summary						
	Cases					
	Valid N	Percent	Missing	Percent	Total N	Percent
I am international; my bike is international	3106	100.0%	0	0.0%	3106	100.0%

I am international*, my bike is international—cross tabulation								
			My bike is international					Total
			Strongly disagree	Disagree	Can't say	Agree	Strongly agree	
I am international	Strongly disagree	Count	9	13	38	33	20	113
		% within I am international	8.0%	11.5%	33.6%	29.2%	17.7%	100.0%
		% within My bike is international	11.7%	3.3%	4.3%	2.8%	3.5%	3.6%
		% of total	0.3%	0.4%	1.2%	1.1%	.6%	3.6%
	Disagree	Count	22	143	237	253	104	759
		% within I am international	2.9%	18.8%	31.2%	33.3%	13.7%	100.0%
		% within My bike is international	28.6%	36.4%	26.5%	21.7%	18.0%	24.4%
		% of Total	0.7%	4.6%	7.6%	8.1%	3.3%	24.4%
	Can't say	Count	20	105	366	354	140	985
		% within I am international	2.0%	10.7%	37.2%	35.9%	14.2%	100.0%
		% within My bike is international	26.0%	26.7%	40.9%	30.4%	24.2%	31.7%
		% of Total	0.6%	3.4%	11.8%	11.4%	4.5%	31.7%
	Agree	Count	19	99	210	454	203	985
		% within I am international	1.9%	10.1%	21.3%	46.1%	20.6%	100.0%
		% within My bike is international	24.7%	25.2%	23.5%	39.0%	35.1%	31.7%
		% of Total	0.6%	3.2%	6.8%	14.6%	6.5%	31.7%
	Strongly agree	Count	7	33	43	70	111	264
		% within I am international	2.7%	12.5%	16.3%	26.5%	42.0%	100.0%
		% within My bike is international	9.1%	8.4%	4.8%	6.0%	19.2%	8.5%
		% of Total	0.2%	1.1%	1.4%	2.3%	3.6%	8.5%
Total		Count	77	393	894	1164	578	3106
		% within I am International	2.5%	12.7%	28.8%	37.5%	18.6%	100.0%
		% within My bike is international	100.0%	100.0%	100.0%	100.0%	100.0%	100.0%
		% of Total	2.5%	12.7%	28.8%	37.5%	18.6%	100.0%

Chi-square tests			
	Value	df	Asymp. Sig. (two-sided)
Pearson chi-square	241.887a	16	0
Likelihood ratio	217.995	16	0
Linear-by-linear association	95.082	1	0
Number of valid cases	3106		

Expected count less than 5. The minimum expected count is 2.80.

TABLE 5.14

Cross Tabulation of Biker–Bike Personality Ratings on the Personality Variable "Extraordinary"

Case Processing Summary						
	Cases					
	Valid N	Percent	Missing	Percent	Total N	Percent
I am extraordinary; my bike is extraordinary	3106	100.0%	0	0.0%	3106	100.0%

I am extraordinary, my bike is extraordinary—cross tabulation								
			My bike is extraordinary					Total
			Strongly disagree	Disagree	Can't say	Agree	Strongly agree	
I am extraordinary	Strongly disagree	Count	20	31	30	14	11	106
		% within I am extraordinary	18.9%	29.2%	28.3%	13.2%	10.4%	100.0%
		% within My bike is extraordinary	16.5%	5.1%	3.6%	1.4%	2.0%	3.4%
		% of Total	0.6%	1.0%	1.0%	.5%	.4%	3.4%
	Disagree	Count	31	239	190	170	47	677
		% within I am extraordinary	4.6%	35.3%	28.1%	25.1%	6.9%	100.0%
		% within My bike is extraordinary	25.6%	39.6%	23.0%	16.7%	8.7%	21.8%
		% of Total	1.0%	7.7%	6.1%	5.5%	1.5%	21.8%
	Can't say	Count	40	191	434	379	167	1211
		% within I am extraordinary	3.3%	15.8%	35.8%	31.3%	13.8%	100.0%
		% within My bike is extraordinary	33.1%	31.6%	52.6%	37.3%	30.9%	39.0%
		% of Total	1.3%	6.1%	14.0%	12.2%	5.4%	39.0%
	Agree	Count	21	117	140	384	190	852
		% within I am extraordinary	2.5%	13.7%	16.4%	45.1%	22.3%	100.0%
		% within My bike is extraordinary	17.4%	19.4%	17.0%	37.8%	35.1%	27.4%
		% of total	0.7%	3.8%	4.5%	12.4%	6.1%	27.4%
	Strongly agree	Count	9	26	31	68	126	260
		% within I am extraordinary	3.5%	10.0%	11.9%	26.2%	48.5%	100.0%
		% within My bike is extraordinary	7.4%	4.3%	3.8%	6.7%	23.3%	8.4%
		% of Total	0.3%	0.8%	1.0%	2.2%	4.1%	8.4%
Total		Count	121	604	825	1015	541	3106
		% within I am extraordinary	3.9%	19.4%	26.6%	32.7%	17.4%	100.0%
		% within My bike is extraordinary	100.0%	100.0%	100.0%	100.0%	100.0%	100.0%
		% of Total	3.9%	19.4%	26.6%	32.7%	17.4%	100.0%

Chi-square tests			
	Value	df	Asymp. Sig. (two-sided)
Pearson chi-square	567.358a	16	0
Likelihood ratio	498.371	16	0
Linear-by-linear association	323.425	1	0
Number of valid cases	3106		
Expected count less than 5. The minimum expected count is 4.13.			

"3—can't say" for rating their own and their bike's personality, respectively. A diverse trend is seen for ratings given for biker–bike personality on this personality variable.

Detailed analysis of the cross-tabulation table reveals that 24.8%, 10.3%, and 14% of the total respondents gave similar high (4—agree or 5—strongly agree), low (1—strongly disagree or 2—disagree), and average (3—can't say) ratings to their own and their bike's personality, respectively. However, 7.9% of the total respondents gave lower ratings (1—strongly disagree or 2—disagree) to themselves and higher ratings (4—agree or 5—strongly agree) to their bikes on this personality variable. And 5.6% respondents gave higher ratings (4—agree or 5—strongly agree) to themselves and lower ratings (1—strongly disagree or 2— disagree) to their bikes. Thus, around 13.5% respondents gave clear incongruent ratings as compared to a majority of congruent ratings given by more than 49.1% (i.e., 1,525 out of 3,106) of the respondents. Also, a large percentage (12.6%) of respondents chose the "3—can't say" option for rating their bikes while attributing higher (4—agree or 5—strongly agree) or lower (1—strongly disagree or 2—disagree) ratings to themselves. Similarly, 25% of the respondents chose the "3—can't say" option for rating themselves while attributing higher (4—agree or 5—strongly agree) or lower (1—strongly disagree or 2—disagree) ratings to their bikes. Except for these 37.4% (12.6% + 25%) of the responses that could not be clearly categorized as congruent or incongruent, the findings showed congruence between the biker–bike personality ratings for 49.1% of the responses.

5.1.15 Synopsis

1. The chi-square significance value is "0" (for 16 degrees of freedom) for all 14 personality variables. This establishes the presence of a significant relationship between the biker and bike personality for each personality variable.
2. A clear trend is observed in the ratings given by the sample population for the seven personality variables "formal, friendly, comfortable, professional, popular, sporty, and sharp." A majority of respondents gave higher ratings (4—agree or 5—strongly agree) to these personality variables.
3. Mixed ratings were given by the sample population for the remaining seven personality variables "heavy, traditional, emotional, mean, aggressive, international, and extraordinary." The sample population seems to have greater incongruence in their own and their bike's personality for these personality variables.
4. It was found that a majority of respondents (Table 5.15) gave congruent ratings to themselves and their bikes for all 14 personality variables. This indicates the congruent relationship between biker–bike personality for the overall sample population.

TABLE 5.15

Percentage of Congruent Responses for Biker–Bike Personalities on 14 Personality Variables

Serial No.	*Emoha* Words	Incongruent (% Responses)	Congruent (% Responses)	Neither Congruent Nor Incongruent (% Responses)	Total %
1	Formal	7.6	**88.1**	4.3	100
2	Friendly	2.2	**84.1**	13.7	100
3	Comfortable	5.4	**85.7**	8.9	100
4	Professional	6	**76.7**	17.3	100
5	Heavy	23.6	**59.3**	17.1	100
6	Traditional	17.5	**54.9**	27.6	100
7	Popular	9.4	**56.1**	34.5	100
8	Emotional	15.9	**45.4**	38.7	100
9	Self-focused	11.2	**53.9**	34.9	100
10	Aggressive	21.3	**49.6**	29.1	100
11	Sporty	17.2	**61.2**	21.6	100
12	Sharp	12	**62.4**	25.6	100
13	International	18.2	**44.8**	37	100
14	Extraordinary	13.5	**49.1**	37.4	100
	Average (%)	12.93	**62.24**	24.84	100

5. Personality variables "heavy and aggressive" had highest percentage of incongruent biker–bike personality rating as compared to other personality variables.

5.1.16 Discussion

The presence of a congruent relationship between the biker–bike personalities for all personality variables was a significant finding of the study. It highlights the preference of the sample population for owning bikes that match their own personality rather than owning bikes with an opposite personality.

However, at this stage, the study does not divide the sample population on the basis of the social cultural segmentation of biking. Even though the results of this section show the relationship between biker–bike personality, it does not investigate the differences in biker–bike personalities for different biking segments. While some personality variables show a clear pattern of ratings for the entire population, there are other variables that have received diverse ratings. The writer felt that the differences in the ratings given by the bikers belonging to different segments should be further investigated. Also, the correlation between the various personality variables required further investigation. The data were therefore further analyzed to study the possibility of clustering the 14 personality variables into biker–bike personality factors.

5.2 Bike Personality Factors

5.2.1 Describing Data with Mean Scores

The reliability of the survey questionnaire was tested using the Cronbach's alpha test (Table 5.16). The Cronbach's alpha rating (0.823) helped establish the reliability of the survey questionnaire.

Table 5.16 shows the mean ratings for the bike personalities reported by 3,106 respondents. Personality variables "friendly, comfortable, popular, formal, and professional" have mean ratings above "4" and can therefore be seen as the dominant bike personality traits among the sample population, whereas the personality variables "sharp, sporty, international, aggressive, traditional, extraordinary, muscular, emotional, and mean" have mean ratings between "3 and 4." These can be seen as neutral bike personality traits among the sample population. The variable "heavy" seems to be the weakest personality trait (mean rating is less than "3").

Figure 5.1 shows the comparison in the bike personality ratings for the five biking segments. The bikes associated with dexterous and exploratory biking and high-esteem biking have higher mean ratings for the personality variables "international, sharp, extraordinary, sporty, aggressive, mean, emotional, and heavy" as compared to bread-and-butter biking, travel-and-comfort biking, and social-networking biking. The latter, on the other hand, have higher ratings for "formal and traditional." Clear differentiation can also be seen among bike personalities associated with dexterous and exploratory biking and high-esteem biking as well as among bread-and-butter biking, travel-and-comfort biking, and social-networking biking.

However, the mean ratings do not explain the correlation between various personality variables. The analysis was therefore followed by testing the correlation between the personality variables.

5.2.2 Parametric Test of Correlation

A correlation matrix for bike personality ratings was prepared to study the correlation between the 14 personality variables (Table 5.17). A correlation value (c) lies between –1 and +1. The significance of correlation value is as follows:

- $0 < c < 0.5$: positive correlation
- $0.5 < c < 1$: strong positive correlation
- $-0.5 < c < 0$: negative correlation
- $-1 < c < -0.5$: strong negative correlation

The eight sets of opposite personality variables: "formal–emotional, traditional–international, heavy–sharp, comfortable–aggressive, professional–sporty,

TABLE 5.16

Mean Ratings for Bike Personalities on Each Personality Variable

Scale: All variables

Case Processing Summary			
		N	%
Cases	Valid	3106	100
	Excluded[a]	0	0
	Total	3106	100
Ion based on all variables in the procedure.			
Reliability statistics			
Cronbach's alpha	**Cronbach's alpha based on standardized items**	**Number of items**	
0.817	0.823	14	
Item statistics			
	Mean	**Std. deviation**	**N**
My bike is friendly	4.35	0.631	3106
My bike is comfortable	4.34	0.714	3106
My bike is popular	4.25	0.828	3106
My bike is formal	4.07	0.836	3106
My bike is professional	4.07	0.819	3106
My bike is sharp	3.71	0.947	3106
My bike is sporty	3.67	1.052	3106
My bike is international	3.57	1.01	3106
My bike is aggressive	3.48	1.081	3106
My bike is traditional	3.45	1.018	3106
My bike is extraordinary	3.4	1.101	3106
My bike is emotional	3.23	0.95	3106
My bike is mean	3.01	1.034	3106
My bike is heavy	2.75	1.088	3106

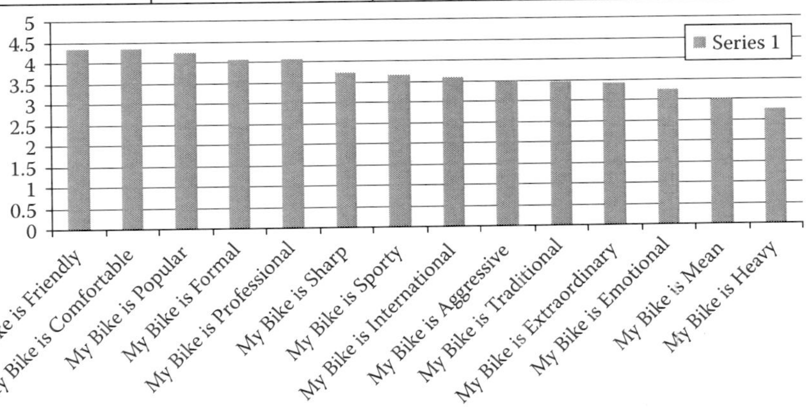

My Bike is:	Bread and Butter Biking	Travel and Comfort Biking	Social Networking Biking	Dextrous and Exploratory	High Esteem Biking	Overall Mean
Formal	4.195	4.216	4.08	3.449	3.34	4.07
Friendly	4.343	4.362	4.333	4.375	4.396	4.35
Comfortable	4.267	4.329	4.338	4.36	4.553	4.34
Professional	3.972	4.133	4.045	3.922	4.068	4.07
Heavy	2.502	2.436	2.856	3.859	3.466	2.75
Traditional	3.526	3.637	3.397	2.926	2.65	3.45
Popular	4.084	4.381	4.182	3.979	4.32	4.25
Emotional	3.167	3.245	3.187	3.304	3.476	3.23
Mean	2.849	3.004	2.926	3.357	3.515	3.01
Aggressive	3.243	3.371	3.434	4.099	4.33	3.48
Sporty	3386	3.475	3.686	4.428	4.699	3.67
Sharp	3.49	3.655	3.682	4.085	4.359	3.71
International	3.41	3.614	3.501	3.629	3.951	3.57
Extraordinary	3.235	3.39	3.33	3.654	4.058	3.4

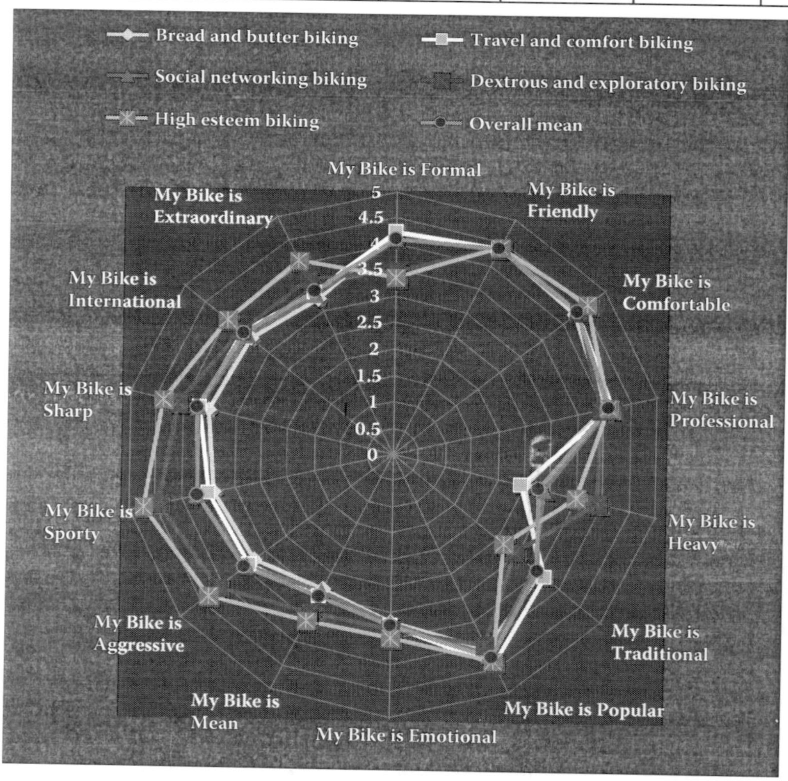

FIGURE 5.1 (See color insert)
Mean ratings for bike personalities associated with each biking segment on each personality variable.

TABLE 5.17
Interitem Correlation Matrix for Bike Personality Ratings on Each Personality Variable

Interitem Correlation Matrix	My Bike Is													
	Formal	Friendly	Comfortable	Professional	Heavy	Traditional	Popular	Emotional	Mean	Aggressive	Sporty	Sharp	International	Extraordinary
My bike is formal	1	0.269	0.181	0.276	–0.052	0.382	0.209	0.093	0.071	–0.013	–0.021	0.066	0.121	0.109
My bike is friendly	0.269	1	0.557	0.464	0.038	0.159	0.307	0.196	0.077	0.268	0.308	0.315	0.339	0.323
My bike is comfortable	0.181	0.557	1	0.463	0.047	0.122	0.327	0.165	0.07	0.273	0.319	0.315	0.349	0.347
My bike is professional	0.276	0.464	0.463	1	0.047	0.211	0.316	0.232	0.095	0.225	0.247	0.278	0.332	0.316
My bike is heavy	–0.052	0.038	0.047	0.047	1	0.063	–0.026	0.193	0.281	0.307	0.27	0.234	0.154	0.227
My bike is traditional	0.382	0.159	0.122	0.211	0.063	1	0.203	0.174	0.18	0.078	0.025	0.107	0.17	0.152
My bike is popular	0.209	0.307	0.327	0.316	–0.026	0.203	1	0.179	0.078	0.175	0.169	0.23	0.343	0.281
My bike is emotional	0.093	0.196	0.165	0.232	0.193	0.174	0.179	1	0.312	0.278	0.203	0.238	0.244	0.261
My bike is mean	0.071	0.077	0.07	0.095	0.281	0.18	0.078	0.312	1	0.318	0.157	0.238	0.188	0.219
My bike is aggressive	–0.013	0.268	0.273	0.225	0.307	0.078	0.175	0.278	0.318	1	0.452	0.447	0.336	0.384
My bike is sporty	–0.021	0.308	0.319	0.247	0.27	0.025	0.169	0.203	0.157	0.452	1	0.551	0.406	0.409
My bike is sharp	0.066	0.315	0.315	0.278	0.234	0.107	0.23	0.238	0.238	0.447	0.551	1	0.442	0.458
My bikc is international	0.121	0.339	0.349	0.332	0.154	0.17	0.343	0.244	0.188	0.336	0.406	0.442	1	0.544
My bike is extraordinary	0.109	0.323	0.347	0.316	0.227	0.152	0.281	0.261	0.219	0.384	0.409	0.458	0.544	1

Continued

TABLE 5.17 (*Continued*)

Interitem Correlation Matrix for Bike Personality Ratings on Each Personality Variable

Item-Total Statistics

	Scale Mean If Item Deleted	Scale Variance If Item Deleted	Corrected Item-Total Correlation	Squared Multiple Correlation	Cronbach's Alpha If Item Deleted
My bike is formal	53.43	60.88	0.209	0.237	0.819
My bike is friendly	53.15	59.347	0.469	0.41	0.807
My bike is comfortable	53.16	58.842	0.452	0.405	0.807
My bike is professional	53.43	58.038	0.448	0.341	0.806
My bike is heavy	54.75	58.019	0.308	0.232	0.815
My bike is traditional	54.05	58.859	0.282	0.217	0.816
My bike is popular	53.25	58.972	0.366	0.225	0.81
My bike is emotional	54.27	57.61	0.402	0.19	0.808
My bike is mean	54.49	57.469	0.368	0.226	0.811
My bike is aggressive	54.02	54.622	0.533	0.373	0.799
My bike is sporty	53.83	55.2	0.512	0.435	0.801
My bike is sharp	53.79	55.203	0.583	0.427	0.797
My bike is international	53.93	54.717	0.574	0.408	0.797
My bike is extraordinary	54.1	53.377	0.604	0.441	0.794

Scale Statistics

Mean	Variance	Std. Deviation	No. of Items
57.5	64.305	8.019	14

popular–extraordinary, and friendly–mean," were studied first. All sets had a positive correlation ($0 < c < 0.5$). This was followed by studying the correlation between individual personality variables.

A strong positive correlation ($0.5 < c < 1$) was seen between "extraordinary and international," "sharp and sporty," and "comfortable and friendly." Strong positive correlation implied that respondents gave distinct congruent ratings to these personality variables while rating their bike. Similarly, higher positive correlation ($0.4 < c < 0.5$) was also seen between "professional and friendly/comfortable," "sporty and aggressive/international/extraordinary," and "sharp and aggressive/international/extraordinary."

No strong negative correlation ($-0.5 < c < 0$) was observed. Some personality variables that had negative correlation ($-0.5 < c < 0$) are "formal with heavy/aggressive/sporty" and "heavy with popular." Thus, while "formal" had positive correlation value with its opposite personality variable "emotional," it had negative correlation with "heavy, aggressive, and sporty." This implies that those respondents who rate their bike as "formal" may not rate their bike as "heavy, aggressive, and sporty" and vice versa.

The study of the correlation matrix indicates correlation between some personality variables. However, the nature and level of correlation was different for different personality variables. The study was therefore followed by a factor analysis (principal component analysis) to investigate the possibility of clustering personality variables into bike personality factors.

5.2.3 Factor Analysis (Principal Component Analysis) with Varimax Rotation Technique with Kaiser Normalization

Table 5.18 shows the results of the KMO (Kaiser–Meyer–Olkin) and Bartlett's test for bike personality ratings. The KMO statistic varies between 0 and 1. A value close to 1 indicates that patterns of correlations are relatively compact and that the factor analysis should yield distinct and reliable factors. Table 5.18 shows that the Kaiser–Meyer–Olkin measure of sampling adequacy is 0.88, which ascertains the reliability of the factor analysis.

Bartlett's measure tests the null hypothesis that the original correlation is an identity matrix. A significant test (i.e., having a significance value less than 0.05) indicates that the R-matrix is not an identity matrix and that there are some relationships between the variables to be included in the factor analysis. Table 5.18 shows that Bartlett's test is highly significant ($p < 0.001$) and therefore confirms that the factor analysis is appropriate.

Table 5.18 shows the communities before and after extraction. Principal component analysis works on the initial assumption that all variance is common; therefore, before extraction the commonalities are all 1. The commonalities in the column labeled "Extraction" reflect the common variance in the data structure.

Table 5.18 also shows the component matrix and rotated component matrix. At this stage, three components (factors) were extracted using SPSS. Both the

TABLE 5.18

Results of Factor Analysis (Principal Component Analysis)

KMO and Bartlett's Test		
Kaiser–Meyer–Olkin measure of sampling adequacy		0.88
Bartlett's test of sphericity	Approx. chi-square	13231.479
	Df	120
	Sig.	0
Communalities		
	Initial	**Extraction**
My bike is formal	1	0.538
My bike is friendly	1	0.58
My bike is comfortable	1	0.589
My bike is professional	1	0.519
My bike is heavy	1	0.454
My bike is traditional	1	0.549
My bike is popular	1	0.375
My bike is emotional	1	0.313
My bike is mean	1	0.491
My bike is aggressive	1	0.516
My bike is sporty	1	0.583
My bike is sharp	1	0.528
My bike is international	1	0.485
My bike is extraordinary	1	0.519
Method: Principal component analysis.		

Component Matrix[a]			
	Component		
	1	2	3
My bike is extraordinary	0.714		
My bike is sharp	0.696		
My bike is international	0.69		
My bike is sporty	0.656		
My bike is aggressive	0.64	−0.325	
My bike is friendly	0.596	0.396	
My bike is comfortable	0.596	0.342	−0.342
My bike is professional	0.568	0.418	
My bike is emotional	0.466		0.305
My bike is popular	0.462	0.402	
My bike is formal		0.614	0.32
My bike is heavy	0.37	−0.504	
My bike is traditional		0.405	0.544
My bike is mean	0.402		0.515
Extraction method: principal component analysis.			
[a] 3 components extracted.			

TABLE 5.18 (*Continued*)
Results of Factor Analysis (Principal Component Analysis)

Rotated Component Matrix[a]			
	Component		
	1	2	3
My bike is aggressive	0.694		
My bike is sporty	0.67	0.311	
My bike is sharp	0.64	0.343	
My bike is heavy	0.623		
My bike is extraordinary	0.586	0.408	
My bike is mean	0.496		0.48
My bike is international	0.495	0.483	
My bike is emotional	0.4		0.366
My bike is friendly		0.742	
My bike is comfortable		0.736	
My bike is professional		0.695	
My bike is popular		0.573	
My bike is traditional			0.697
My bike is formal		0.445	0.546
Extraction method: principal component analysis. Rotation method: Varimax with Kaiser normalization.			
[a] Rotation converged in eight iterations.			
Component transformation matrix			
Component	1	2	3
1	0.75	0.625	0.219
2	−0.66	0.679	0.322
3	0.053	−0.386	0.921
Extraction method: principal component analysis. Rotation method: Varimax with Kaiser normalization.			

matrices show the factor loadings for each bike personality variable onto each factor. On the basis of the higher factor loadings, the following three bike personality factors were identified:

Bike personality factor 1

My bike is INTERNATIONAL
My bike is EXTRAORDINARY
My bike is SHARP
My bike is SPORTY
My bike is AGGRESSIVE
My bike is HEAVY
My bike is MEAN
My bike is EMOTIONAL

Bike personality factor 2

My bike is FRIENDLY

My bike is COMFORTABLE

My bike is PROFESSIONAL

My bike is POPULAR

Bike personality factor 3

My bike is TRADITIONAL

My bike is FORMAL

Figure 5.2 shows the visual representation of the different bike personality factors in a radar chart arrangement. As is clear from the chart, all opposite personality variables other than "heavy–sharp" are grouped in different factors. The previous correlation results showed that the opposite words had positive correlation, but the results of the factor analysis show that the level of correlation was not significant enough to group these words together.

These three bike personality factors can be seen as three personality dimensions on which the sample biker population can be distinguished. The personality variables associated with these factors are correlated. These bike personality factors represent a framework for comparison of bike personalities associated with different biking segments.

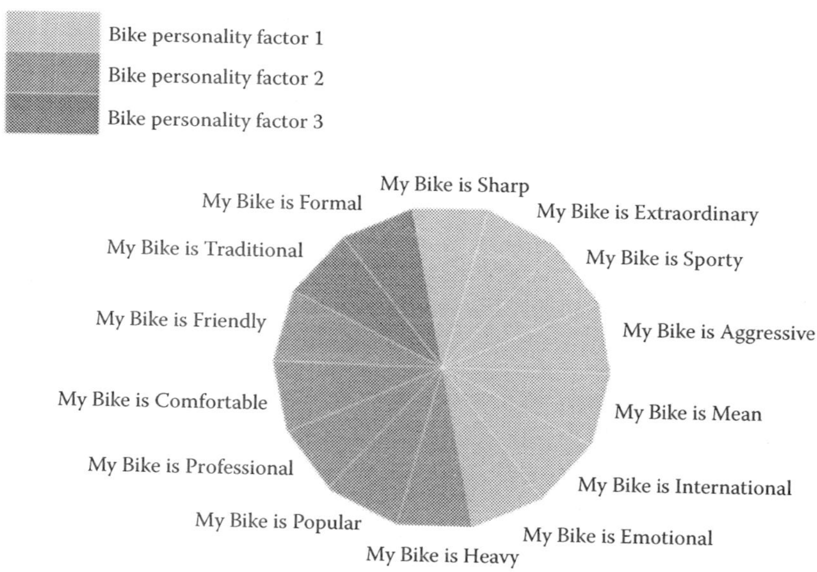

FIGURE 5.2 (See color insert)
Bike personality factors.

5.3 Biker Personality Factors

5.3.1 Describing Data with Mean Scores

The reliability of the survey questionnaire was tested using the Cronbach's alpha test (Table 5.19). Cronbach's alpha rating (0.0675) establishes the reliability of the survey questionnaire.

Table 5.19 shows the mean ratings for the biker personalities of the 3,106 bikers who participated in the survey. Biker personality variables "friendly, professional, comfortable, and formal and professional" have mean ratings above "4" and can therefore be seen as the dominant biker personality traits among the sample population. On the other hand, personality variables "sharp, emotional, sporty, popular, traditional, aggressive, international, and extraordinary" have mean ratings between "3 and 4" and can therefore be seen as neutral personality traits of the sample population. "Mean and heavy" seem to be the weak personality traits as their mean score is less than "3."

5.3.2 Parametric Test of Correlation

The correlation matrix for the biker personality ratings was prepared to study the correlation between the 14 personality variables (Table 5.20).

The eight sets of opposite personality variables, "formal—emotional, traditional—international, heavy—sharp, comfortable—aggressive, professional—sporty, popular—extraordinary, and friendly—mean," were analyzed first. A negative correlation ($-0.5 < c < 0$) was observed between "mean—friendly" and between "aggressive—comfortable" as well. All other sets have a positive correlation ($0 < c < 0.5$). This was followed by studying the correlation between individual personality variables.

Neither a strong positive correlation ($0.5 < c < 1$) nor a strong negative correlation ($-0.5 < c < 0$) was seen between the biker personalities reported on the 14 personality variables. Most of the variables thus seem to have a positive correlation. Some sets of personality variables that have a positive correlation value ($c > 0.2$) are as follows: "professional with comfortable/friendly/popular/sharp," "international with popular/sports/sharp/extraordinary," "sharp with sports/popular," and "mean with aggressive."

The analysis of the correlation matrix indicated correlation between some personality variables. However, the nature and level of correlation was different for different personality variables. The test of correlation was therefore followed by a factor analysis (principal component analysis) to investigate the possibility of clustering the personality variables into biker personality factors.

TABLE 5.19

Mean Ratings for Biker Personalities on Each Personality Variable

Scale: all variables

Case Processing Summary			
		N	%
Cases	Valid	3106	100
	Excluded	0	0
	Total	3106	100
List wise deletion based on all variables in the procedure.			
Reliability Statistics			
Cronbach's alpha	**Cronbach's alpha based on standardized items**	**No. of items**	
0.675	0.683	14	
Item statistics			
	Mean	**Std. deviation**	**N**
I am friendly	4.48	0.577	3106
I am professional	4.27	0.718	3106
I am comfortable	4.2	0.684	3106
I am formal	4.16	0.695	3106
I am sharp	3.93	0.733	3106
I am emotional	3.78	0.957	3106
I am sporty	3.74	0.943	3106
I am popular	3.64	0.834	3106
I am traditional	3.48	1.022	3106
I am aggressive	3.21	1.127	3106
I am international	3.17	1.009	3106
I am extraordinary	3.16	0.969	3106
I am mean	2.86	1.178	3106
I am heavy	2.56	1.104	3106

Biker Self Personality – All Survey Responses

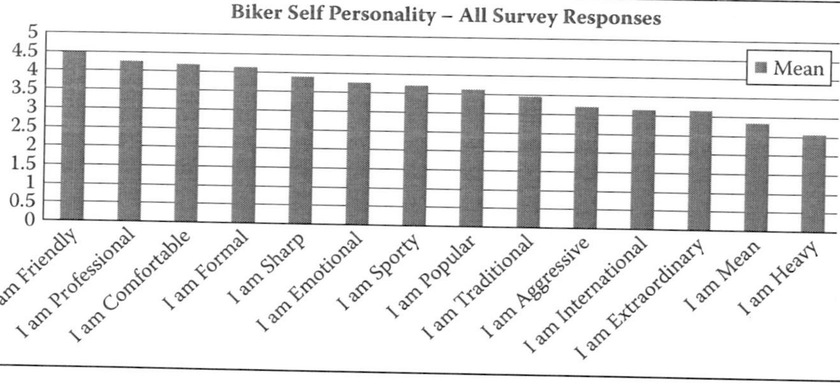

TABLE 5.20

Interitem Correlation Matrix for Biker Personality Ratings on Each Personality Variable

	Interitem Correlation Matrix													
	I Am													
	Formal	Friendly	Comfortable	Professional	Heavy	Traditional	Popular	Emotional	Mean	Aggressive	Sporty	Sharp	International	Extraordinary
I am formal	1	0.108	0.092	0.221	0.053	0.157	0.122	0.063	0.064	0.017	0.067	0.158	0.061	0.094
I am friendly	0.108	1	0.246	0.16	0.008	0.013	0.214	0.098	−0.037	0.001	0.183	0.168	0.09	0.094
I am comfortable	0.092	0.246	1	0.16	0.062	0.092	0.163	0.009	0.031	−0.026	0.178	0.122	0.13	0.13
I am professional	0.221	0.16	0.16	1	0.075	0.075	0.202	0.025	0.019	0.034	0.121	0.234	0.193	0.143
I am heavy	0.053	0.008	0.062	0.075	1	0.198	0.083	0.072	0.195	0.161	0	0.031	0.103	0.124
I am traditional	0.157	0.013	0.092	0.075	0.198	1	0.055	0.122	0.188	0.117	0.061	0.068	0.018	0.089
I am popular	0.122	0.214	0.163	0.202	0.083	0.055	1	0.137	0.011	0.088	0.16	0.246	0.282	0.282
I am emotional	0.063	0.098	0.009	0.025	0.072	0.122	0.137	1	0.035	0.16	0.049	0.105	0.077	0.099
I am mean	0.064	−0.037	0.031	0.019	0.195	0.188	0.011	0.035	1	0.306	0.026	0.074	0.045	0.123
I am aggressive	0.017	0.001	−0.026	0.034	0.161	0.117	0.088	0.16	0.306	1	0.142	0.143	0.11	0.19
I am sporty	0.067	0.183	0.178	0.121	0	0.061	0.16	0.049	0.026	0.142	1	0.248	0.212	0.196
I am sharp	0.158	0.168	0.122	0.234	0.031	0.068	0.246	0.105	0.074	0.143	0.248	1	0.247	0.29
I am international	0.061	0.09	0.13	0.193	0.103	0.018	0.282	0.077	0.045	0.11	0.212	0.247	1	0.389
I am extraordinary	0.094	0.094	0.13	0.143	0.124	0.089	0.282	0.099	0.123	0.19	0.196	0.29	0.389	1

Continued

TABLE 5.20 (*Continued*)

Interitem Correlation Matrix for Biker Personality Ratings on Each Personality Variable

Item-Total Statistics

	Scale Mean If Item Deleted	Scale Variance If Item Deleted	Corrected Item-Total Correlation	Squared Multiple Correlation	Cronbach's Alpha If Item
I am formal	52.17	35.739	0.213	0.092	0.667
I am friendly	51.85	36.373	0.185	0.13	0.67
I am comfortable	52.13	35.836	0.207	0.116	0.668
I am professional	52.06	35.336	0.251	0.137	0.664
I am heavy	53.77	33.686	0.241	0.115	0.666
I am traditional	52.85	33.904	0.256	0.116	0.663
I am popular	52.69	34.037	0.335	0.186	0.654
I am emotional	52.55	35.152	0.17	0.065	0.674
I am mean	53.47	32.974	0.269	0.179	0.663
I am aggressive	53.12	32.757	0.308	0.168	0.656
I am sporty	52.59	33.924	0.29	0.168	0.658
I am sharp	52.4	34.34	0.363	0.194	0.652
I am international	53.16	32.863	0.356	0.231	0.649
I am extraordinary	53.17	32.214	0.44	0.268	0.638

Scale Statistics

Mean	Variance	Std. deviation	No. of items
56.33	37.996	6.164	14

5.3.3 Factor Analysis (Principal Component Analysis) with Varimax Rotation Technique with Kaiser Normalization

Table 5.21 shows the results of the KMO (Kaiser–Meyer–Olkin) and Bartlett's test for biker personality ratings. The KMO statistic varies between 0 and 1. A value close to 1 indicates that patterns of correlations are relatively compact and that the factor analysis should yield distinct and reliable factors. Table 5.18 shows that the Kaiser–Meyer–Olkin measure of sampling adequacy is 0.772, which ascertains the reliability of the factor analysis.

Bartlett's measure tests the null hypothesis that the original correlation is an identity matrix. A significant test (i.e., having a significance value less than 0.05) indicates that the R-matrix is not an identity matrix and that there are some relationships between the variables to be included in the factor analysis. Table 5.21 shows that Bartlett's test is highly significant ($p < 0.001$) and therefore confirms that the factor analysis is appropriate.

Table 5.21 shows the communities before and after extraction. Principal component analysis works on the initial assumption that all variance is common; therefore, before extraction the commonalities are all 1. The commonalities in the column labeled "Extraction" reflect the common variance in the data structure.

Table 5.21 also shows the component matrix and rotated component matrix. At this stage, five components (factors) were extracted using SPSS. Both the matrices show the factor loadings for each biker personality variable onto each factor. Based on the higher factor loadings, the following three biker personality factors were identified:

Biker personality factor 1

- I am INTERNATIONAL
- I am EXTRAORDINARY
- I am SHARP
- I am SPORTY
- I am POPULAR

Biker personality factor 2

- I am MEAN
- I am TRADITIONAL
- I am HEAVY
- I am AGGRESSIVE

Biker personality factor 3

- I am COMFORTABLE
- I am FRIENDLY

TABLE 5.21

Results of Factor Analysis (Principal Component Analysis)

KMO and Bartlett's Test		
Kaiser–Meyer–Olkin measure of sampling adequacy		0.772
Bartlett's Test of Sphericity	Approx. chi-square	5,114.092
	Df	120
	Sig.	0
Communalities		
	Initial	**Extraction**
I am formal	1	0.611
I am friendly	1	0.473
I am comfortable	1	0.58
I am professional	1	0.442
I am heavy	1	0.414
I am traditional	1	0.485
I am popular	1	0.397
I am emotional	1	0.764
I am mean	1	0.501
I am aggressive	1	0.483
I am sporty	1	0.362
I am sharp	1	0.428
I am international	1	0.499
I am extraordinary	1	0.521
Method: Principal component analysis.		

Component Matrix[a]					
	Component				
	1	2	3	4	5
I am extraordinary	0.628				
I am international	0.567				
I am sharp	0.56				
I am popular	0.536				
I am sporty	0.48				
I am professional	0.423				
I am mean		0.62			
I am aggressive		0.465			
I am friendly		−0.426			
I am formal			0.544		−0.426
I am traditional			0.493		
I am emotional				0.809	
I am comfortable					0.492
I am heavy					
Extraction method: principal component analysis.					
[a] 5 components extracted.					

TABLE 5.21 (*Continued*)

Results of Factor Analysis (Principal Component Analysis)

Rotated Component Matrix[a]					
	Component				
	1	2	3	4	5
I am international	0.704				
I am extraordinary	0.7				
I am sharp	0.536				
I am sporty	0.5				
I am popular	0.481				
I am mean		0.682			
I am traditional		0.589			
I am heavy		0.588			
I am aggressive		0.467			
I am comfortable			0.703		
I am friendly			0.626		
I am formal				0.773	
I am professional				0.589	
I am emotional					0.865

Extraction Method: Principal Component analysis. Rotation method: Varimax with Kaiser normalization.

[a] Rotation converged in seven iterations.

Component Transformation Matrix					
Component	1	2	3	4	5
1	0.791	0.39	0.275	0.341	0.176
2	−0.15	0.815	−0.531	−0.174	0.033
3	−0.575	0.284	0.387	0.659	0.073
4	−0.117	−0.051	0.093	−0.241	0.958
5	−0.092	0.317	0.696	−0.601	−0.214

Extraction method: Principal component analysis. Rotation method: Varimax with Kaiser normalization.

Biker personality factor 4

I am FORMAL

I am PROFESSIONAL

Biker personality factor 5

I am EMOTIONAL

These five biker personality factors can be seen as five personality dimensions on which the sample biker population can be distinguished. The variables associated with these factors are correlated. These biker personality factors (Figure 5.3) represent a framework for comparing the biker personalities associated with different biking segments.

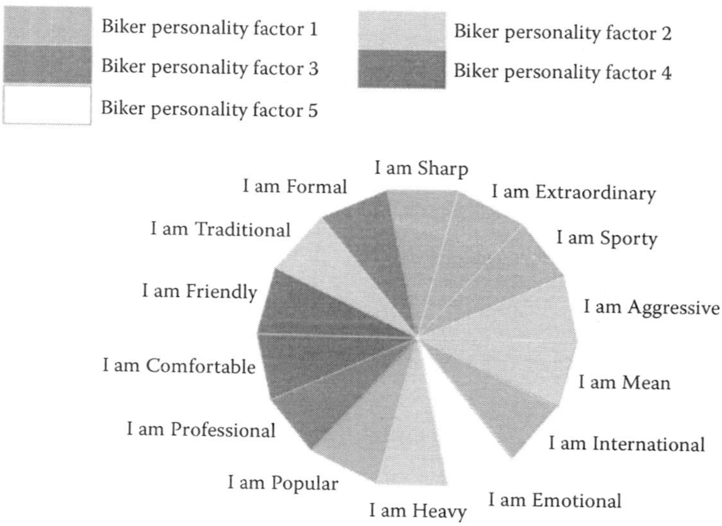

FIGURE 5.3 (See color insert)
Biker personality factors.

5.4 Comparing Bike Personality of Biking Segments

One-way ANOVA was conducted between the identified bike personality factors and the bikes belonging to the five biking segments. The objective of the test was to investigate the presence of a significant difference, if any, in the personality of bikes belonging to the different biking segments.

Table 5.22 shows the results of the ANOVA analysis. It was found that all the three bike personality factors have a significance value of "0" (for 4 degrees of freedom), which implies a significant difference between the five groups (biking segments). The biking segments were compared on each of these factors.

5.4.1 Comparing Biking Segments on Bike Personality Factor 1

Table 5.22 shows the comparison between the bike personalities of the five biking segments for bike personality factor 1 (international, extraordinary, sharp, sporty, aggressive, heavy, mean, and emotional). It was found that the five biking segments could be clearly differentiated (significance value < 0.05) on this bike personality factor. The mean difference between each biking segments was analyzed to arrange the biking segments in order of their ratings for this factor.

It was found that the bikes belonging to high-esteem biking received significantly higher ratings for bike personality factor 1 as compared to bikes belonging to bread-and-butter biking (mean difference: 1.62, sign. value: 0),

TABLE 5.22

Comparing Biking Segments on Bike Personality Factors

ANOVA						
		Sum of Squares	df	Mean Square	F	Sig.
REGR factor score Bike 1	Between groups	614.247	4	153.562	191.185	0
	Within groups	2490.753	3101	0.803		
	Total	3105	3105			
REGR factor score Bike 2	Between groups	96.931	4	24.233	24.981	0
	Within groups	3008.069	3101	0.97		
	Total	3105	3105			
REGR factor score Bike 3	Between groups	244.133	4	61.033	66.156	0
	Within groups	2860.867	3101	0.923		
	Total	3105	3105			

Post Hoc Tests							
Multiple Comparisons							
LSD							
						95% confidence interval	
Dependent variable	(I) AD	(J) AD	Mean difference (I–J)	Std. error	Sig.	Lower bound	Upper bound
REGR factor score Bike 1	Bread-and-butter biking	Travel and comfort	−0.09756561	0.06149172	0.113	−0.2181342	0.023003
		Social networking biking	−0.32783043*	0.06276114	0	−0.450888	−0.2047728
		Dexterous and exploratory biking	0.07770608	0	−1.5785997	−1.2738786	
		High-esteem biking	−1.62448634*	0.10487225	0	−1.8301124	−1.4188602
	Travel and comfort	Bread-and-butter biking	0.09756561	0.06149172	0.113	−0.023003	0.2181342
		Social networking biking	−0.23026481*	0.03633341	0	−0.3015048	−0.1590248
		Dexterous and exploratory biking	−1.32867354*	0.05847555	0	−1.4433283	−1.2140188

Continued

TABLE 5.22 (*Continued*)

Comparing Biking Segments on Bike Personality Factors

Post Hoc Tests							
Multiple Comparisons							
LSD							
						95% confidence interval	
Dependent variable	(I) AD	(J) AD	Mean difference (I–J)	Std. error	Sig.	Lower bound	Upper bound
	Social networking biking	Bread-and-butter biking	.32783043*	0.06276114	0	0.2047728	0.450888
		Travel and comfort	.23026481*	0.03633341	0	0.1590248	0.3015048
		Dexterous and exploratory biking	−1.09840873*	0.05980903	0	−1.215678	−0.9811394
		High-esteem biking	−1.29665591*	0.09239629	0	−1.47782	−1.1154918
	Dexterous and exploratory biking	Bread-and- butter biking	1.42623915*	0.07770608	0	1.2738786	1.5785997
		Travel and comfort	1.32867354*	0.05847555	0	1.2140188	1.4433283
		Social networking biking	1.09840873*	0.05980903	0	0.9811394	1.215678
		High-esteem biking	−0.19824719	0.10313267	0.055	−0.4004624	0.0039681
	High-esteem biking	Bread-and-butter biking	1.62448634*	0.10487225	0	1.4188602	1.8301124
		Travel and comfort	1.52692072*	0.09153876	0	1.347438	1.7064035
		Social networking biking	1.29665591*	0.09239629	0	1.1154918	1.47782
		Dexterous and exploratory biking	0.19824719	0.10313267	0.055	−0.0039681	0.4004624
REGR factor score Bike 2	Bread-and-butter biking	Travel and comfort	−0.18787117[a]	0.06757643	0.005	−0.3203703	−0.0553721
		Social networking biking	−0.00589706	0.06897147	0.932	−0.1411314	0.1293373
		Dexterous and exploratory biking	0.43013780[a]	0.08539522	0	0.2627009	0.5975747
		High-esteem biking	0.11629209	0.11524954	0.313	−0.1096811	0.3422652

	Travel and comfort	Bread-and-butter biking	0.18787117[a]	0.06757643	0.005	0.0553721	0.3203703
		Social networking biking	0.18197410[a]	0.03992866	0	0.1036848	0.2602634
		Dexterous and exploratory biking	0.61800897[a]	0.0642618	0	0.492009	0.744009
		High-esteem biking	0.30416326[a]	0.10059668	0.003	0.1069204	0.5014061
	Social networking biking	Bread-and-butter biking	0.00589706	0.06897147	0.932	−0.1293373	0.1411314
		Travel and comfort	−0.18197410[a]	0.03992866	0	−0.2602634	−0.1036848
		Dexterous and exploratory biking	0.43603486[a]	0.06572723	0	0.3071616	0.5649082
		High-esteem biking	0.12218915	0.10153906	0.229	−0.0769015	0.3212798
	Dexterous and exploratory biking	Bread-and-butter biking	−0.43013780[a]	0.08539522	0	−0.5975747	−0.2627009
		Travel and comfort	−0.61800897[a]	0.0642618	0	−0.744009	−0.492009
		Social networking biking	−0.43603486[a]	0.06572723	0	−0.5649082	−0.3071616
		High-esteem biking	−0.31384571[a]	0.11333782	0.006	−0.5360705	−0.0916209
	High-esteem biking	Bread-and-butter biking	−0.11629209	0.11524954	0.313	−0.3422652	0.1096811
		Travel and comfort	−0.30416326[a]	0.10059668	0.003	−0.5014061	−0.1069204
		Social networking biking	−0.12218915	0.10153906	0.229	−0.3212798	0.0769015
		Dexterous and exploratory biking	0.31384571[a]	0.11333782	0.006	0.0916209	0.5360705
REGR factor score Bike 3	Bread-and-butter biking	Travel and comfort	−0.11977226	0.06590224	0.069	−0.2489887	0.0094442
		Social networking biking	0.19976389[a]	0.06726272	0.003	0.0678799	0.3316479
		Dexterous and exploratory biking	0.67875287[a]	0.08327958	0	0.5154642	0.8420416
		High-esteem biking	0.92151955[a]	0.11239426	0	0.7011448	1.1418943
	Travel and comfort	Bread-and-butter biking	0.11977226	0.06590224	0.069	−0.0094442	0.2489887
		Social networking biking	0.31953614[a]	0.03893944	0	0.2431864	0.3958858
		Dexterous and exploratory biking	0.79852512[a]	0.06266973	0	0.6756467	0.9214035
		High-esteem biking	1.04129181[a]	0.09810442	0	0.8489356	1.233648

Continued

TABLE 5.22 (*Continued*)

Comparing Biking Segments on Bike Personality Factors

Post Hoc Tests							
Multiple Comparisons							
LSD							
						95% confidence interval	
Dependent variable	(I) AD	(J) AD	Mean difference (I–J)	Std. error	Sig.	Lower bound	Upper bound
		Travel and comfort	−0.31953614[a]	0.03893944	0	−0.3958858	−0.2431864
		Dexterous and exploratory biking	0.47898898[a]	0.06409885	0	0.3533085	0.6046695
		High-esteem biking	0.72175566[a]	0.09902345	0	0.5275975	0.9159139
	Dexterous and exploratory biking	Bread-and butter biking	−0.67875287[a]	0.08327958	0	−0.8420416	−0.5154642
		Travel and comfort	−0.79852512[a]	0.06266973	0	−0.9214035	−0.6756467
		Social networking biking	−0.47898898[a]	0.06409885	0	−0.6046695	−0.3533085
		High-esteem biking	.24276668[a]	0.11052991	0.028	0.0260475	0.4594859
	High-esteem biking	Bread-and-butter biking	−0.92151955[a]	0.11239426	0	−1.1418943	−0.7011448
		Travel and comfort	−1.04129181[a]	0.09810442	0	−1.233648	−0.8489356
		Social networking biking	−0.72175566[a]	0.09902345	0	−0.9159139	−0.5275975
		Dexterous and exploratory biking	−0.24276668[a]	0.11052991	0.028	−0.4594859	−0.0260475

[a] The mean difference is significant at the 0.05 level.

travel and comfort biking (mean difference: 1.52, sign. value: 0), and social networking biking (mean difference: 1.29, sign. value: 0). Similarly, the bikes belonging to dexterous and exploratory biking received significantly higher ratings for this bike personality factor as compared to bikes belonging to bread-and-butter biking (mean difference: 1.42, sign. value: 0), travel-and-comfort biking (mean difference: 1.32, sign. value: 0), and social-networking biking (mean difference: 1.09, sign. value: 0). Also, the bikes belonging to social-networking biking received significantly higher ratings for this bike personality factor as compared to bikes belonging to bread-and-butter biking (mean difference: 0.32, sign. value: 0) and travel and comfort biking (mean difference: 0.23, sign. value: 0). The highest mean difference was observed between bikes belonging to high-esteem biking and bread-and-butter biking (mean difference: 1.62448634*, sign. value:0)

However, no significant difference was observed between the bikes belonging to high-esteem biking and dexterous and exploratory biking (sign. value = 0.055 > 0.05) as well as between the bikes belonging bread-and-butter biking and travel-and-comfort biking (sign. value = 0.113 > 0.05)

Figure 5.4 shows the comparison between the five biking segments on the basis of bike personality factor 1. It is clear from Table 5.22 and Figure 5.4 that the factor value increases as we move up the biking pyramid from bread-and-butter biking to high-esteem biking. This implies that the bikes belonging to

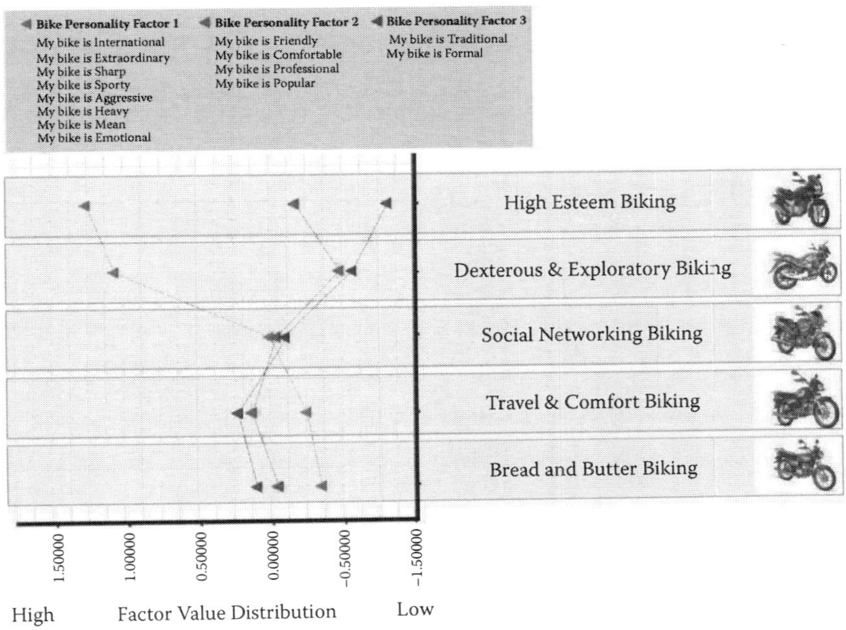

FIGURE 5.4 (See color insert)
Multi–scatter plot showing the bike personality associated with different biking segments.

high-esteem biking and dexterous and exploratory biking are seen as the most "international, extraordinary, sharp, sporty, aggressive, heavy, mean, and emotional," whereas the bikes belonging to bread-and-butter biking and travel-and-comfort biking score least on these personality variables.

5.4.2 Comparing Biking Segments on Bike Personality Factor 2

Table 5.22 shows the comparison between the bike personalities of the five biking segments for bike personality factor 2 (friendly, comfortable, professional, and popular). It was found that the five biking segments could be clearly differentiated (significance value < 0.05) on this factor. The mean difference between each biking segments was analyzed to arrange the biking segments in order of their ratings for this factor.

It was found that the bikes belonging to travel and comfort biking received significantly higher ratings for this bike personality factor as compared to bikes belonging to bread-and-butter biking (mean difference: 0.187, sign. value: 0.005), social networking biking (mean difference: 0.181, sign. value: 0), dexterous and exploratory biking (mean difference: 0.618, sign. value: 0), and high-esteem biking (mean difference: 0.304, sign. value: 0.003). On the other hand, the bikes belonging to dexterous and exploratory biking received significantly lower ratings for this bike personality factor as compared to bikes belonging to bread-and-butter biking (mean difference: −4.301, sign. value: 0), travel-and-comfort biking (mean difference: −0.618, sign. value: 0), social-networking biking (mean difference: −0.436, sign. value: 0), and high-esteem biking (mean difference: −0.313, sign. value: 0.006). The highest mean difference was observed between bikes belonging to travel and comfort biking and dexterous and exploratory biking (mean difference: 0.61800897, sign. value: 0.001).

However, no significant difference was observed between the bikes belonging to bread-and-butter biking and social-networking biking (sign. value = 0.932 > 0.05) as well as between bikes belonging to bread-and-butter biking and high-esteem biking (sign. value = 0.313 > 0.05).

Figure 5.4 shows the comparison between the five biking segments on the basis of bike personality factor 2. As is clear from Table 5.22 and Figure 5.4, the factor value increases as we move up the biking pyramid from travel and comfort biking to dexterous and exploratory biking. This implies that the bikes belonging to travel-and-comfort biking are seen as the most "friendly, comfortable, professional, and popular," whereas the bikes belonging to dexterous and exploratory biking score least on these personality variables.

The travel-and-comfort biking is the most popular biking segment in India. Also, the highest percentage of respondents in this survey belongs to this biking segment. The clear preference of these respondents toward bikes with bike personality factor 2 represents an important finding in this study. Also, unlike bike personality factor 1, this personality factor differentiates between the personalities of bikes belonging to high-esteem biking and dexterous and exploratory biking.

5.4.3 Comparing Biking Segments on Bike Personality Factor 3

Table 5.22 shows the comparison between the bike personalities of the five biking segments for bike personality factor 3 (formal and traditional). It was found that the five biking segments could be clearly differentiated (significance value < 0.05) on this factor. The mean difference between each biking segment was analyzed to arrange the biking segments in order of their ratings for this factor.

It was found that the bikes belonging to travel and comfort biking received significantly higher ratings for this bike personality factor as compared to bikes belonging to social-networking biking (mean difference: 1.29, sign. value: 0), dexterous and exploratory biking (mean difference: 1.29, sign. value: 0), and high-esteem biking (mean difference: 1.29, sign. value: 0). Similarly, the bikes belonging to social-networking biking received significantly higher ratings for this bike personality factor as compared to bikes belonging to dexterous and exploratory biking (mean difference: 1.29, sign. value: 0) and high-esteem biking (mean difference: 1.29, sign. value: 0). Also, the bikes belonging to dexterous and exploratory biking received significantly higher ratings for this bike personality factor as compared to bikes belonging to high-esteem biking (mean difference: 1.29, sign. value: 0). The highest mean difference was observed between bikes belonging to travel and comfort biking and high-esteem biking (mean difference: 1.04129181, sign. value: 0.001).

However, no significant difference was observed between the bikes belonging to bread-and-butter biking and travel-and-comfort biking (sign. value = 0.069 > 0.05).

Figure 5.4 shows the comparison between the five biking segments on the basis of bike personality factor 3. As is clear from the Figure 5.4, the factor value decreases as we move up the biking pyramid from bread and butter biking to high esteem biking. This implies that the bikes belonging to bread and butter biking and travel and comfort biking are seen as the most "formal" and "traditional," whereas the bikes belonging to higher levels of biking pyramid score least on these personality variables. The preference for bike personality factor 3 is just the reverse of that for bike personality factor 1. This highlights an incongruent relationship between these two bike personality factors.

5.5 Comparing Biker Personality of Biking Segments

The one-way ANOVA method was applied between the identified biker personality factors and the bikers belonging to five biking segments. The objective of the test was to investigate the presence of a significant difference, if any, in the personalities of bikers belonging to different biking segments.

Table 5.23 shows the results of the ANOVA analysis. The biker personality factors 1, 3, and 4 have a significance value of "0" (with 4 degrees of freedom), which implies a strong significant difference between the five groups (biking segments), whereas biker personality factor 5 has a significance value of 0.04 (<0.05), which implies a lesser significant difference between the five biking segments. On the other hand, biker personality factor 2 has a significance value of 0.189 (>0.05), which implies that the different biking segments do not significantly differ on this particular factor. The biking segments were compared on each of the identified factors.

5.5.1 Comparing Biking Segments on Biker Personality Factor 1

Table 5.23 shows the comparison between the biker personalities of the five biking segments for biker personality factor 1 (international, extraordinary, sharp, sporty, and popular). It was found that the five biking segments could be clearly differentiated (significance value < 0.05) on this personality factor. The mean difference between each biking segment was analyzed to arrange the biking segments in order of their ratings for this factor.

It was found that the bikers belonging to high-esteem biking received significantly higher ratings for this personality factor as compared to bikers belonging to bread-and-butter biking (mean difference: 0.461, sign. value: 0), travel-and-comfort biking (mean difference: 0.463, sign. value: 0), and social-networking biking (mean difference: 0.456, sign. value: 0). Similarly, the bikers belonging to dexterous and exploratory biking received significantly higher ratings for this personality factor as compared to bikers belonging to bread-and-butter biking (mean difference: 0.281, sign. value: 0.001), travel-and-comfort biking (mean difference: 0.283, sign. value: 0), and social networking biking (mean difference: 0.275, sign. value: 0). The highest mean difference was observed between bikers belonging to high-esteem biking and travel-and-comfort biking (mean difference: 1.62448634, sign. value: 0).

However, no significant difference was observed between the bikers belonging to high-esteem biking and dexterous and exploratory biking (sign. value = 0.115 > 0.05). Similarly, no significant difference was observed between the bikers belonging to bread-and-butter biking and travel-and-comfort biking (sign. value = 0.977 > 0.05) as well as between bikers belonging to bread-and-butter biking and social-networking biking (sign. value = 0.937 > 0.05).

Figure 5.5 shows the comparison between the five biking segments on the basis of biker personality factor 1. It is clear from Table 5.23 and Figure 5.5 that the factor value increases as we move up the biking pyramid from bread-and-butter biking to high-esteem biking. This implies that the bikers belonging to high-esteem biking and dexterous and exploratory biking rate themselves as "international, extraordinary, sharp, sporty, and popular," whereas the bikers belonging to bread-and-butter biking, travel-and-comfort biking, and social-networking biking score least on these personality variables.

TABLE 5.23
Comparing Biking Segments on Biker Personality Factors

ANOVA						
		Sum of Squares	df	Mean Square	F	Sig.
REGR factor score 1	Between groups	38.887	4	9.722	9.832	0
	Within groups	3066.113	3101	0.989		
	Total	3105	3105			
REGR factor score 2	Between groups	6.134	4	1.534	1.535	0.189
	Within groups	3098.866	3101	0.999		
	Total	3105	3105			
REGR factor score 3	Between groups	25.468	4	6.367	6.411	0
	Within groups	3079.532	3101	0.993		
	Total	3105	3105			
REGR factor score 4	Between groups	60.569	4	15.142	15.424	0
	Within groups	3044.431	3101	0.982		
	Total	3105	3105			
REGR factor score 5	Between groups	9.991	4	2.498	2.503	0.04
	Within groups	3095.009	3101	0.998		
	Total	3105	3105			

Continued

TABLE 5.23 (*Continued*)

Comparing Biking Segments on Biker Personality Factors

Post hoc tests							
Multiple Comparisons							
LSD							
Dependent variable	(I) Motor bike segment?	(J) Motor bike segment?	Mean difference	Std. error	Sig.	95% confidence	
						Lower bound	Upper bound
REGR factor score 1	Bread-and-butter biking	Travel and comfort	0.00200177	0.06822529	0.977	−0.1317696	0.1357731
		Social networking biking	−0.00549462	0.06963372	0.937	−0.1420275	0.1310383
		Dexterous and exploratory biking	−0.28120799[a]	0.08621518	0.001	−0.4502526	−0.1121634
		High-esteem biking	−0.46150671[a]	0.11635615	0	−0.6896496	−0.2333638
	Travel and comfort	Bread-and-butter biking	−0.00200177	0.06822529	0.977	−0.1357731	0.1317696
		Social networking biking	−0.00749639	0.04031206	0.852	−0.0865374	0.0715446
		Dexterous and exploratory biking	−0.28320977[a]	0.06487884	0	−0.4104196	−0.1559999
		High-esteem biking	−0.46350848[a]	0.10156259	0	−0.6626452	−0.2643717
	Social networking biking	Bread-and- butter biking	0.00549462	0.06963372	0.937	−0.1310383	0.1420275
		Travel and comfort	0.00749639	0.04031206	0.852	−0.0715446	0.0865374
		Dexterous and exploratory biking	−0.27571337[a]	0.06635834	0	−0.4058241	−0.1456026
		High-esteem biking	−0.45601209[a]	0.10251403	0	−0.6570143	−0.2550098
	Dexterous and exploratory biking	Bread-and- butter biking	0.28120799[a]	0.08621518	0.001	0.1121634	0.4502526
		Travel and comfort	0.28320977[a]	0.06487884	0	0.1559999	0.4104196
		Social networking biking	0.27571337[a]	0.06635834	0	0.1456026	0.4058241
		High-esteem biking	−0.18029872	0.11442608	0.115	−0.4046573	0.0440598

	High-esteem biking	Bread-and-butter biking	0.46150671[a]	0.11635615	0	0.2333638	0.6896496
		Travel and comfort	0.46350848[a]	0.10156259	0	0.2643717	0.6626452
		Social networking biking	0.45601209[a]	0.10251403	0	0.2550098	0.6570143
		Dexterous and exploratory biking	0.18029872	0.11442608	0.115	−0.0440598	0.4046573
REGR factor score 2	Bread-and-butter biking	Travel and comfort	−0.03973039	0.06858872	0.562	−0.1742143	0.0947535
		Social networking biking	−0.04630338	0.07000466	0.508	−0.1835636	0.0909568
		Dexterous and exploratory biking	0.01476159	0.08667444	0.865	−0.1551835	0.1847067
		High-esteem biking	0.18955824	0.11697597	0.105	−0.0398	0.4189165
	Travel and comfort	Bread-and-butter biking	0.03973039	0.06858872	0.562	−0.0947535	0.1742143
		Social networking biking	−0.00657298	0.04052679	0.871	−0.0860351	0.0728891
		Dexterous and exploratory biking	0.05449198	0.06522444	0.404	−0.0733955	0.1823795
		High-esteem biking	0.22928864[a]	0.10210361	0.025	0.0290911	0.4294862
	Social networking biking	Bread-and-butter biking	0.04630338	0.07000466	0.508	−0.0909568	0.1835636
		Travel and comfort	0.00657298	0.04052679	0.871	−0.0728891	0.0860351
		Dexterous and exploratory biking	0.06106496	0.06671182	0.36	−0.0697389	0.1918688
		High-esteem biking	0.23586162[a]	0.10306011	0.022	0.0337886	0.4379346
	Dexterous and exploratory biking	Bread-and-butter biking	−0.01476159	0.08667444	0.865	−0.1847067	0.1551835
		Travel and comfort	−0.05449198	0.06522444	0.404	−0.1823795	0.0733955
		Social networking biking	−0.06106496	0.06671182	0.36	−0.1918688	0.0697389
		High-esteem biking	0.17479666	0.11503562	0.129	−0.050757	0.4003504
	High-esteem biking	Bread-and-butter biking	−0.18955824	0.11697597	0.105	−0.4189165	0.0398
		Travel and comfort	−0.22928864[a]	0.10210361	0.025	−0.4294862	−0.0290911
		Social networking biking	−0.23586162[a]	0.10306011	0.022	−0.4379346	−0.0337886
		Dexterous and exploratory biking	−0.17479666	0.11503562	0.129	−0.4003504	0.050757

Continued

TABLE 5.23 ***(Continued)***

Comparing Biking Segments on Biker Personality Factors

Post hoc tests							
Multiple Comparisons							
LSD							
Dependent variable	(I) Motor bike segment?	(J) Motor bike segment?	Mean difference	Std. error	Sig.	95% confidence Lower bound	Upper bound
		Social networking biking	−0.17892317[a]	0.06978594	0.01	−0.3157545	−0.0420918
		Dexterous and exploratory biking	−0.33111554[a]	0.08640364	0	−0.5005297	−0.1617014
		High-esteem biking	−0.40581950[a]	0.1166105	0.001	−0.6344611	−0.1771779
	Travel and comfort	Bread-and-butter biking	0.10413285	0.06837443	0.128	−0.0299309	0.2381966
		Social networking biking	−0.07479032	0.04040018	0.064	−0.1540041	0.0044235
		Dexterous and exploratory biking	−0.22698268[a]	0.06502066	0	−0.3544706	−0.0994948
		High-esteem biking	−0.30168665[a]	0.10178461	0.003	−0.5012587	−0.1021146
	Social networking biking	Bread-and-butter biking	0.17892317[a]	0.06978594	0.01	0.0420918	0.3157545
		Travel and comfort	0.07479032	0.04040018	0.064	−0.0044235	0.1540041
		Dexterous and exploratory biking	−0.15219237[a]	0.06650339	0.022	−0.2825875	−0.0217972
		High-esteem biking	−0.22689633[a]	0.10273812	0.027	−0.428338	−0.0254547
	Dexterous and exploratory biking	Bread-and-butter biking	0.33111554[a]	0.08640364	0	0.1617014	0.5005297
		Travel and comfort	0.22698268[a]	0.06502066	0	0.0994948	0.3544706
		Social networking biking	0.15219237[a]	0.06650339	0.022	0.0217972	0.2825875
		High-esteem biking	−0.07470397	0.11467621	0.515	−0.299553	0.150145

	High-esteem biking	Bread-and-butter biking	0.40581950[a]	0.1166105	0.001	0.1771779	0.6344611
		Travel and comfort	0.30168665[a]	0.10178461	0.003	0.1021146	0.5012587
		Social networking biking	0.22689633[a]	0.10273812	0.027	0.0254547	0.428338
		Dexterous and exploratory biking	0.07470397	0.11467621	0.515	−0.150145	0.299553
REGR factor score 4	Bread-and-butter biking	Travel and comfort	0.00134764	0.06798364	0.984	−0.1319499	0.1346452
		Social networking biking	0.16294527[a]	0.06938708	0.019	0.026896	0.2989946
		Dexterous and exploratory biking	0.23390554[a]	0.08590981	0.007	0.0654597	0.4023514
		High-esteem biking	0.68488954[a]	0.11594402	0	0.4575547	0.9122244
	Travel and comfort	Bread-and-butter biking	−0.00134764	0.06798364	0.984	−0.1346452	0.1319499
		Social networking biking	0.16159763[a]	0.04016927	0	0.0828366	0.2403587
		Dexterous and exploratory biking	0.23255790[a]	0.06464904	0	0.1057986	0.3593172
		High-esteem biking	0.68354190[a]	0.10120286	0	0.4851105	0.8819733
	Social networking biking	Bread-and-butter biking	−0.16294527[a]	0.06938708	0.019	−0.2989946	−0.026896
		Travel and comfort	−0.16159763[a]	0.04016927	0	−0.2403587	−0.0828366
		Dexterous and exploratory biking	0.07096027	0.0661233	0.283	−0.0586896	0.2006102
		High-esteem biking	0.52194427[a]	0.10215093	0	0.321654	0.7222346
	Dexterous and exploratory biking	Bread-and-butter biking	−0.23390554[a]	0.08590981	0.007	−0.4023514	−0.0654597
		Travel and comfort	−0.23255790[a]	0.06464904	0	−0.3593172	−0.1057986
		Social networking biking	−0.07096027	0.0661233	0.283	−0.2006102	0.0586896
		High-esteem biking	0.45098400[a]	0.11402079	0	0.2274201	0.6745479
	High-esteem biking	Bread-and- butter biking	−0.68488954[a]	0.11594402	0	−0.9122244	−0.4575547
		Travel and comfort	−0.68354190[a]	0.10120286	0	−0.8819733	−0.4851105
		Social networking biking	−0.52194427[a]	0.10215093	0	−0.7222346	−0.321654
		Dexterous and exploratory biking	−0.45098400[a]	0.11402079	0	−0.6745479	−0.2274201

Continued

TABLE 5.23 ***(Continued)***

Comparing Biking Segments on Biker Personality Factors

Post hoc tests							
Multiple Comparisons							
LSD							
Dependent variable	(I) Motor bike segment?	(J) Motor bike segment?	Mean difference	Std. error	Sig.	95% confidence Lower bound	95% confidence Upper bound
		Social networking biking	−0.00596772	0.06996107	0.932	−0.1431425	0.131207
		Dexterous and exploratory biking	−0.19157191[a]	0.08662048	0.027	−0.3614112	−0.0217326
		High-esteem biking	−0.13614731	0.11690315	0.244	−0.3653627	0.0930681
	Travel and comfort	Bread-and-butter biking	0.07752579	0.06854602	0.258	−0.0568744	0.211926
		Social networking biking	0.07155807	0.04050156	0.077	−0.0078545	0.1509707
		Dexterous and exploratory biking	−0.11404612	0.06518384	0.08	−0.241854	0.0137617
		High-esteem biking	−0.05862152	0.10204005	0.566	−0.2586944	0.1414514
	Social networking biking	Bread-and-butter biking	0.00596772	0.06996107	0.932	−0.131207	0.1431425
		Travel and comfort	−0.07155807	0.04050156	0.077	−0.1509707	0.0078545
		Dexterous and exploratory biking	−0.18560419[a]	0.06667029	0.005	−0.3163266	−0.0548818
		High-esteem biking	−0.13017958	0.10299595	0.206	−0.3321268	0.0717676
	Dexterous and exploratory biking	Bread-and-butter biking	0.19157191[a]	0.08662048	0.027	0.0217326	0.3614112
		Travel and comfort	0.11404612	0.06518384	0.08	−0.0137617	0.241854
		Social networking biking	0.18560419[a]	0.06667029	0.005	0.0548818	0.3163266
		High-esteem biking	0.05542461	0.114964	0.63	−0.1699887	0.2808379
	High-esteem biking	Bread-and-butter biking	0.13614731	0.11690315	0.244	−0.0930681	0.3653627
		Travel and comfort	0.05862152	0.10204005	0.566	−0.1414514	0.2586944
		Social networking biking	0.13017958	0.10299595	0.206	−0.0717676	0.3321268
		Dexterous and exploratory biking	−0.05542461	0.114964	0.63	−0.2808379	0.1699887

[a] The mean difference is significant at the 0.05 level.

5.5.2 Comparing Biking Segments on Biker Personality Factor 2

Table 5.23 shows the comparison between the biker personalities of the five biking segments for biker personality factor 2 (mean, traditional, heavy, and aggressive). It was found that all biking segments could not be clearly differentiated (significance value: 0.189 > 0.05) on this personality factor. However, the mean difference between each biking segment was analyzed to arrange the biking segments in order of their ratings for this factor.

It was found that only bikers belonging to high-esteem biking received significantly lower ratings for this personality factor as compared to bikers belonging to travel-and-comfort biking (mean difference: −0.229, sign. value: 0.025) and social-networking biking (mean difference: −0.235, sign. value: 0.022). However, no significant difference was observed between the bikers belonging to other biking segments (sign. value > 0.05). The highest mean difference was observed between bikers belonging to high-esteem biking and travel-and-comfort biking (mean difference: 0.61800897, sign. value: 0.001).

Figure 5.5 shows the comparison between the five biking segments on the basis of biker personality factor 2. It is clear from Table 5.23 and Figure 5.5 that the factor value does not follow any clear trend as we move up the biking pyramid from bread-and-butter biking to high-esteem biking. This implies that the bikers belonging to high-esteem biking and dexterous and exploratory

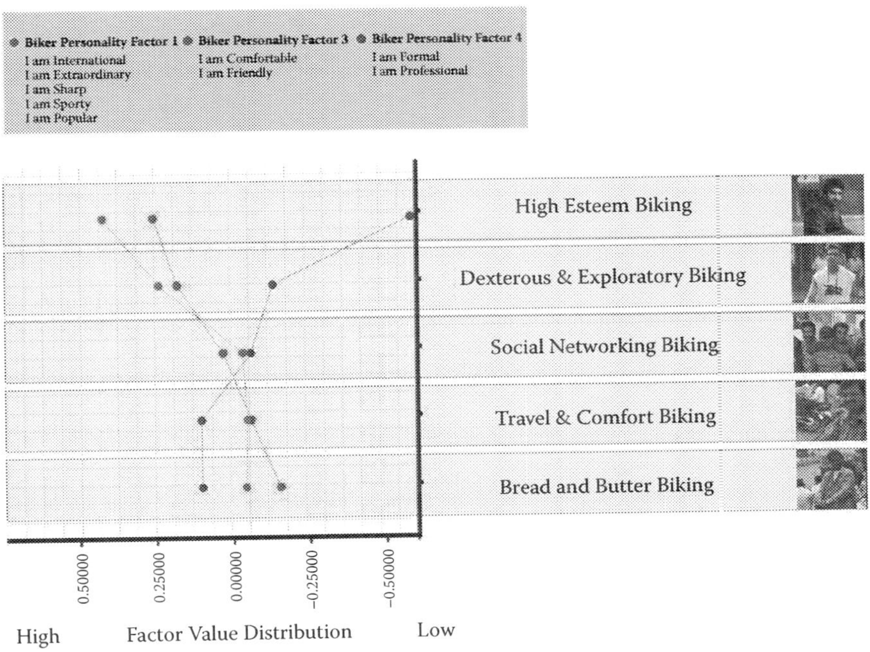

FIGURE 5.5 (See color insert)
Multi-scatter plot showing the biker personality associated with different biking segments.

biking rate themselves higher on personality variables such as "mean, traditional, heavy, aggressive" as compared to the bikers belonging to high-esteem biking, bread-and-butter biking, and dexterous and exploratory biking.

5.5.3 Comparing Biking Segments on Biker Personality Factor 3

Table 5.23 shows the comparison between the biker personalities of the five biking segments for biker personality factor 3 (comfortable and friendly). It was found that the five biking segments could be clearly differentiated (significance value < 0.05) on this personality factor. The mean difference between each biking segment was analyzed to arrange the biking segments in order of their ratings for this factor.

It was found that the bikers belonging to high-esteem biking received significantly higher ratings for this personality factor as compared to bikers belonging to bread-and-butter biking (mean difference: 0.405, sign. value: 0.001), travel-and-comfort biking (mean difference: 0.301, sign. value: 0.003), and social networking biking (mean difference: 0.226, sign. value: 0.027). Similarly, the bikers belonging to dexterous and exploratory biking received significantly higher ratings for this personality factor as compared to bikers of bread-and-butter biking (mean difference: 0.331, sign. value: 0), travel-and-comfort biking (mean difference: 0.226, sign. value: 0), and social-networking biking (mean difference: 0.152, sign. value: 0.022). Also, social-networking biking was higher than bread-and-butter biking (mean difference: 0.178, sign. value: 0.01). The highest mean difference was observed between bikers belonging to high-esteem biking and bread-and-butter biking (mean difference: 1.04129181, sign. value: 0.001)

However, no significant difference was observed between the bikers belonging to high esteem biking and dexterous and exploratory biking (sign. value = 0.515 > 0.05). Similarly, no significant difference was observed between the bikers belonging to travel-and-comfort biking and bread-and-butter biking (sign. value = 0.128 > 0.05) as well as between bikers belonging to travel-and-comfort biking and social-networking biking (sign. value = 0.064 > 0.05).

Figure 5.5 shows the comparison between the five biking segments on the basis of biker personality factor 3. It is clear from Table 5.23 and Figure 5.5 that the factor value increases as we move up the biking pyramid from bread-and-butter biking to high-esteem biking. This implies that the bikers belonging to high-esteem biking and dexterous and exploratory biking rate themselves as "friendly and comfortable," whereas the bikers belonging to bread-and-butter biking score least on these personality variables.

5.5.4 Comparing Biking Segments on Biker Personality Factor 4

Table 5.23 shows the comparison between the biker personalities of the five biking segments for biker personality factor 4 (formal and professional).

It was found that the five biking segments could be clearly differentiated (significance value < 0.05) on this personality factor. The mean difference between each biking segment was analyzed to arrange the biking segments in order of their ratings for this factor.

It was found that the bikers belonging to bread-and-butter biking received significantly higher ratings for this personality factor as compared to bikers belonging to social-networking biking (mean difference: 0.162, sign. value: 0.019), dexterous and exploratory biking (mean difference: 0.233, sign. value: 0.007), and high-esteem biking (mean difference: 0.684, sign. value: 0). Similarly, the bikers belonging to travel-and-comfort biking received significantly higher ratings for this personality factor as compared to bikers belonging to social networking biking (mean difference: 0.161, sign. value: 0), dexterous and exploratory biking (mean difference: 0.232, sign. value: 0), and high-esteem biking (mean difference: 0.683, sign. value: 0). Also, high-esteem biking received significantly lower ratings as compared to social-networking biking (mean difference: −0.521, sign. value: 0) and dexterous and exploratory biking (mean difference: −0.450, sign. value: 0). The highest mean difference was observed between bikers belonging to high-esteem biking and bread-and-butter biking (mean difference: 0.68488954, sign. value: 0).

However, no significant difference was observed between the bikers belonging to bread-and-butter biking and travel-and-comfort biking (sign. value = 0.984 > 0.05). Similarly, no significant difference was observed between the bikers belonging to social-networking biking and bikers of dexterous and exploratory biking (sign. value = 0.283 > 0.05).

Figure 5.5 shows the comparison between the five biking segments on the basis of biker personality factor 4. It is clear from Table 5.23 and Figure 5.5 that the factor value decreases as we move up the biking pyramid from bread-and-butter biking to high-esteem biking. This implies that the bikers belonging to bread-and-butter biking and travel-and-comfort biking segment rate themselves as "formal and professional," whereas the bikers belonging to high-esteem biking score least on these personality variables.

5.5.5 Comparing Biking Segments on Biker Personality Factor 5

Table 5.23 shows the comparison between the biker personalities of the five biking segments for biker personality factor 5 (emotional). It was found that the five biking segments could not be as clearly differentiated (significance value is 0.04 < 0.05) on this personality factor. The mean difference between each biking segment was analyzed to arrange the biking segments in order of their ratings for this factor.

It was found that only bikers belonging to dexterous and exploratory biking received significantly higher ratings for this personality factor as compared to bikers belonging to bread-and-butter biking (mean difference: 0.191, sign. value: 0.027) and social-networking biking (mean difference: 0.185, sign. value: 0.005). No significant difference was seen among bikers belonging to other biking segments. The highest mean difference was observed between bikers belonging to dexterous and exploratory biking and bread-and-butter biking (mean difference: 1.04129181, sign. value: 0.001).

Figure 5.5 shows the comparison between the five biking segments on the basis of personality factor 5. It is clear from the Table 5.23 and the Figure 5.5 that biker personality factor 5 is highest for bikers belonging to dexterous and exploratory biking. This implies that the bikers belonging to dexterous and exploratory biking are distinctively more "emotional" as compared to other bikers in the sample population.

5.6 Emotive Quality of Biking Segments

The previous section compares the biker and bike personalities of the five biking segments on the five biker personality factors and three bike personality factors, respectively. As explained earlier in the book, it is this person–product personality that together gives an overall emotive quality to a product experience. Therefore, the writer impelled to study the emotive quality of biking segments with respect to the biker–bike personality factors. The results of the ANOVA analysis reveal that all five biking segments can be clearly distinguished on all three bike personality factors and three (out of five) biker personality factors.

The review of all six personality factors reveals a resemblance in the nature of three sets of personality factors. A similarity was seen among bike personality factor 1 and biker personality factor 1, bike personality factor 2 and biker personality factor 3, and bike personality factor 3 and biker personality factor 4 on the basis of the related personality variables. Personality variables "international, extraordinary, sporty, and sharp" are common to bike personality factor 1 and biker personality factor 1. Similarly, personality variables "comfortable and friendly" are common to bike personality factor 2 and biker personality factor 3. And personality variable "formal" is common to bike personality factor 3 and biker personality factor 4. Because of the resemblance in the nature of these three sets of biker and bike personality factors, they were given the same color code (green for bike personality factor 1 and biker personality factor 1, red for bike personality factor 2 and biker personality factor 3, and blue for bike personality factor 3 and biker personality factor 4). The multi–scatter plot shows that these three sets of personality factors also share a congruent relationship in factor value distribution across all five biking segments.

Figure 5.6 shows the multi–scatter plot for comparing the emotive quality of the five biking segments on these six significant personality factors. The figure illustrates the difference in the emotive quality of the different biking segments.

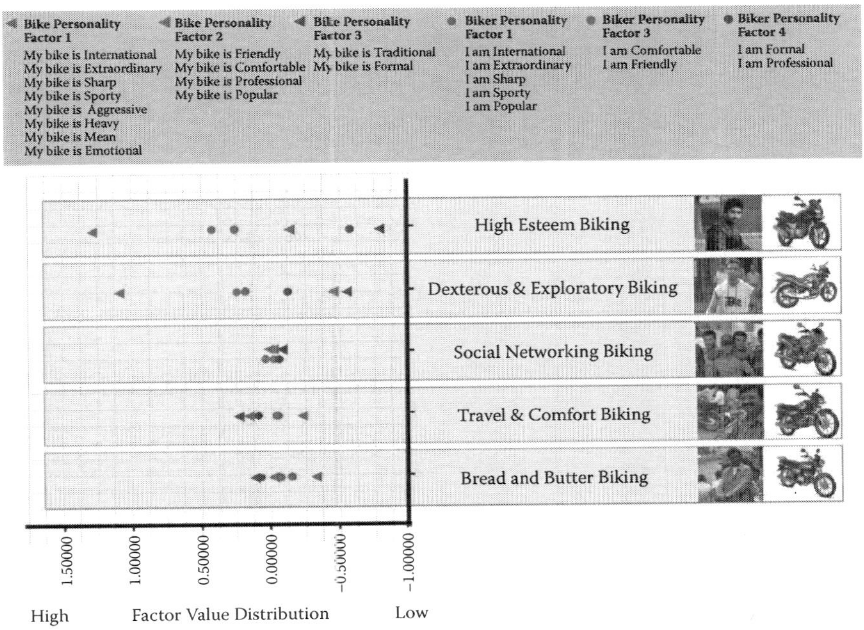

FIGURE 5.6 (See color insert)
Multi–scatter plot showing the emotive quality of biking segments in India.

5.6.1 Emotive Quality of High-Esteem Biking

1. The emotive quality of high-esteem biking is characterized by a high factor value for bike personality factor 1 and biker personality factor 1 followed by bike personality factor 2 and biker personality factor 3 and least factor value for bike personality factor 3 and biker personality factor 4.
2. The emotive quality of this biking segment is characterized by a widely spread factor value distribution between the most dominant factor (bike personality factor 1) and the least dominant factor (bike personality factor 3).
3. The factor value for bike personality factor 1 is the highest for this biking segment, while the factor value for biker personality factor 4 is the lowest among all segments.

5.6.2 Emotive Quality of Dexterous and Exploratory Biking

1. Similar to high-esteem biking, the emotive quality of dexterous and exploratory biking is also characterized by a high factor value for bike personality factor 1 and biker personality factor 1 and the least factor value for bike personality factor 3.

2. However, the factor value distribution between the most dominant factor (bike personality factor 1) and the least dominant factor (bike personality factor 3) is less widely spread as compared to the case of high esteem biking.
3. The factor value of biker personality factor 4 is higher, whereas bike personality factor 2 is lower in the case of this biking segment as compared to the high-esteem biking segment.

5.6.3 Emotive Quality of Social-Networking Biking

1. The emotive quality of social-networking biking is the most interesting in comparison to other biking segments. The factor value distribution between the most dominant factor (biker personality factor 3) and the least dominant factor (bike personality factor 3) is the least in this case. All the six personality factors have similar factor values, thereby signifying a balanced emotive quality. The distinct emotive quality of this segment clearly differentiates it from the other four biking segments.

5.6.4 Emotive Quality of Travel-and-Comfort Biking

1. The emotive quality of travel and comfort biking correlates inversely with that of dexterous and exploratory biking. The emotive quality of this biking segment is characterized by a high factor value for bike personality factor 3 and biker personality factor 4 followed by bike personality factor 2 and biker personality factor 3 and the least factor value for biker personality factor 1 and bike personality factor 1.
2. The factor value distribution between the most dominant factor (bike personality factor 3) and the least dominant factor (bike personality factor 1) is more widely spread as compared to the case of social networking biking.

5.6.5 Emotive Quality of Bread-and-Butter Biking

1. The emotive quality of bread-and-butter biking correlates highly with travel-and-comfort biking and inversely correlates to high-esteem biking. Similar to bread-and-butter biking, the emotive quality of this biking segment is characterized by a high factor value for bike personality factor 3 and biker personality factor 4 followed by bike personality factor 2 and biker personality factor 3 and the least factor value for biker personality factor 1 and bike personality factor 1.
2. The most dominant personality factor (bike personality factor 3) and the least dominant factor (bike personality factor 1) are the same for both travel and comfort biking and bread-and-butter biking.

However, the values of these factors are lower in the case of bread-and-butter biking.

The results show that each biking segment has a distinct emotive quality that differentiates it from other biking segments. However, while there is diversity in the emotive quality of different biking segments, yet there is a congruent relationship between similar *biker–bike personality factors* for individual biking segments. This was also established in Section 5.1, where it was observed that while there was diversity in the response associated with personality variables, yet there was a congruence in the individual responses related to biker–bike personality on each personality variable. This explains that it is, in fact, the similar difference in bike as well as biker personality that distinguishes the emotive quality of the five biking segments.

These results lead to the formulation of an emotion-centered research framework for product styling, which is the core focus of the book.

6

Emotion-Centered Research Framework for Design Innovation

Development of an emotion-centered research framework was the final objective of this study and the result of the research undertaken. The final chapter of the book confirms the research premise and proposes an emotion-centered research framework for product styling.

6.1 Salient Findings of the Study

The results of the qualitative and quantitative analysis establish the difference in the emotive needs and emotive quality associated with each of these biking segments. The summary of the results of the qualitative and quantitative analysis is presented in the following text.

6.1.1 Emotive Needs Associated with Different Biking Segments

The survey generated more than 10,000 responses to the question, "Why do you ride a motorbike? Give three reasons." The content of each response was analyzed in detail. A total of 33 unique need categories were identified by the content analysis technique. The results of the qualitative content analysis (Chapter 4) established the differentiation in the emotive needs of the selected sample of 3,106 bikers belonging to different biking segments. It was found that the bread-and-butter biking segment is associated with needs for "fuel efficiency," "economy," and "low maintenance." Biking is a necessity rather than a choice for these customers. The rational needs such as fuel efficiency and economy dominate over the need for an "affective experience." On the other hand, "comfort," "affective experience," and "saving time" represent the needs for travel-and-comfort biking. The higher importance of the needs for "accessibility" and "commuting" differentiate social-networking biking from travel-and-comfort biking. For bikers belonging to the dexterous and exploratory biking as well as high-esteem biking segments, biking is primarily an "affective experience." While "looks and style" is the key need for dexterous and exploratory biking, the need to "enhance personality" and to acquire "personal freedom" is important for high-esteem biking. Figure 6.1

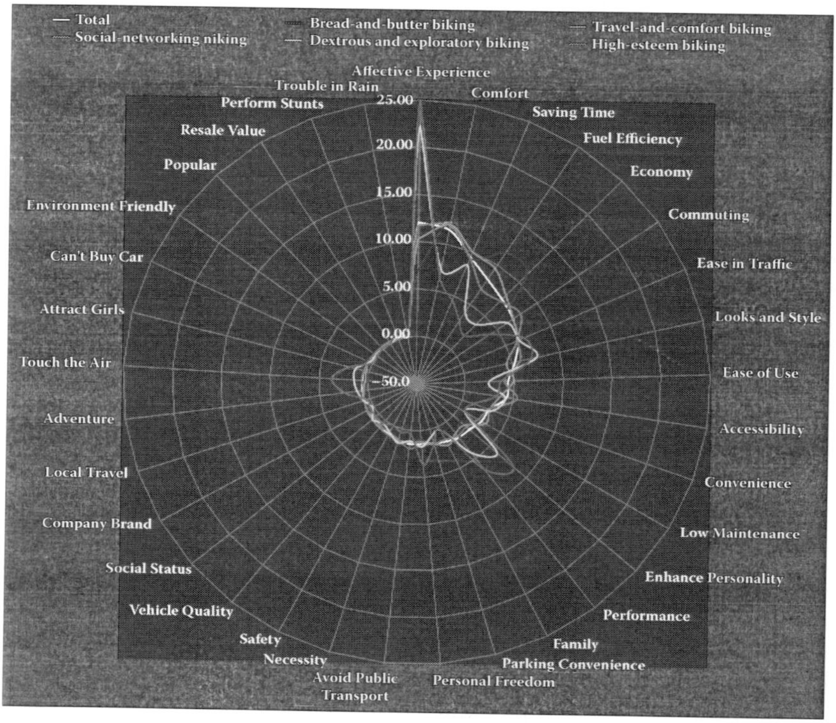

FIGURE 6.1 (See color insert)
Comparison between biking needs for each biking segment.

(same as Figure 4.2 discussed in Section 4.7, Chapter 4) shows the differentiation between biking needs associated with different biking segments.

Furthermore, the mapping of various biking needs on the H-E-A-T diagram (Das, 2002) established the differentiation in the quality of biking needs associated with each biking segment (discussed in Section 4.8, Chapter 4). And the mapping of the respondents' affective experiences on the circumplex of product emotions [developed by Russell (1980) and Watson and Tellgen (1985) and adapted by Desmet (2002)] established the differentiation in the emotive intensity of biking needs associated with each biking segment (discussed in Section 4.9, Chapter 4).

The results thereby ascertain the social-cultural segmentation of motorbike ownership experiences in India in the form of five biking segments: bread-and-butter biking, travel-and-comfort biking, social-networking biking, dexterous and exploratory biking, and high-esteem biking.

6.1.2 Emotive Quality of Different Biking Segments

A biker–bike personality measurement instrument was developed (discussed in Section 3.1, Chapter 3) by identifying a cluster of 14 adjectives (formal, emotional, professional, sporty, extraordinary, popular, mean, friendly, aggressive, comfortable, sharp, heavy, international, and traditional) that can act as personality variables relevant to motorbikes in India. The developed instrument was used to investigate the biker–bike personality of 3,106 motorbike owners of 23 different motorbikes through an online survey conducted in different states of India.

The biker and bike personality was separately investigated on each identified personality variable (Section 5.1, Chapter 5). A subset of five biker and three bike personality factors was identified by grouping of correlated personality variables using the method of factor analysis (Sections 5.2 and 5.3, Chapter 5). Eight personality factors were identified as biker personality factor 1 (I am international, I am extraordinary, I am sharp, I am sporty, I am popular), biker personality factor 2 (I am mean, I am traditional, I am heavy, I am aggressive), biker personality factor 3 (I am comfortable, I am friendly), biker personality factor 4 (I am formal, I am professional), biker personality factor 5 (I am emotional), bike personality factor 1 (My bike is international, My bike is extraordinary, My bike is sharp, My bike is sporty, My bike is aggressive, My bike is heavy, My bike is mean, My bike is emotional), bike personality factor 2 (My bike is friendly, My bike is comfortable, My bike is professional, My bike is popular), and bike personality factor 3 (My bike is traditional, My bike is formal).

The analysis of biking segments with respect to these factors revealed a significant differentiation among the five biking segments on six out of the eight personality factors (discussed in Sections 5.4 and 5.5, Chapter 5). The emotive quality of all biking segments was mapped on these six personality factors. The emotive quality of each biking segment is visually presented in Figure 6.2 (same as Figure 5.6 discussed in Section 5.6, Chapter 5). The results of the study establish the difference in the emotive quality of different biking segments.

The research proposed that the bread-and-butter biking, travel-and-comfort biking, social-networking biking, dexterous and exploratory biking, and high-esteem biking segments can be differentiated on the basis of the associated emotive needs for biking and emotive quality of biking. The results of the qualitative and quantitative analysis empirically establish the difference in emotive needs and emotive quality of each biking segment. The results thereby confirm the research premise.

6.2 Emotion-Centered Research Framework

The findings of the study were correlated and represented in form of an emotion-centered model for motorbike styling in India as shown in Figure 6.3. The developed model divides motorbike ownership experiences in India into five biking segments and illustrates the differences in the emotive needs

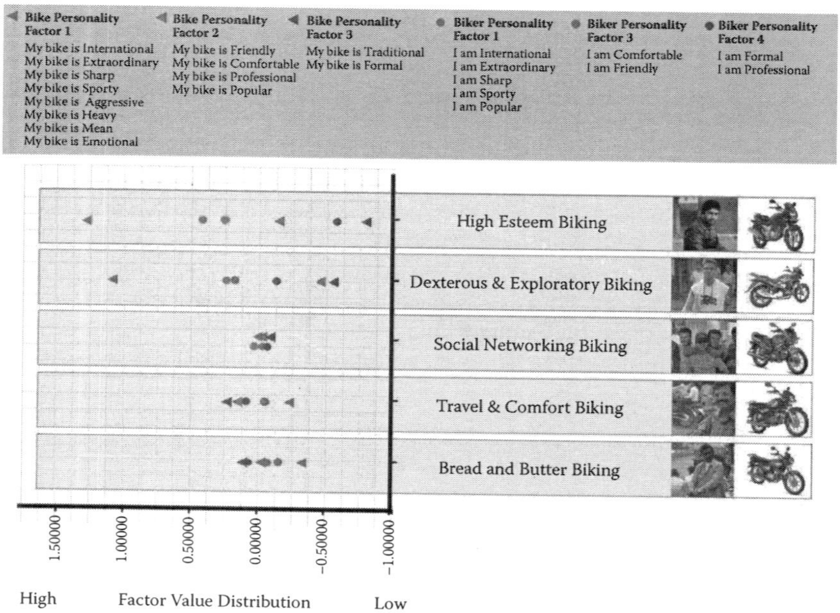

FIGURE 6.2 (See color insert)
Multi–scatter plot showing the emotive quality of biking segments in India.

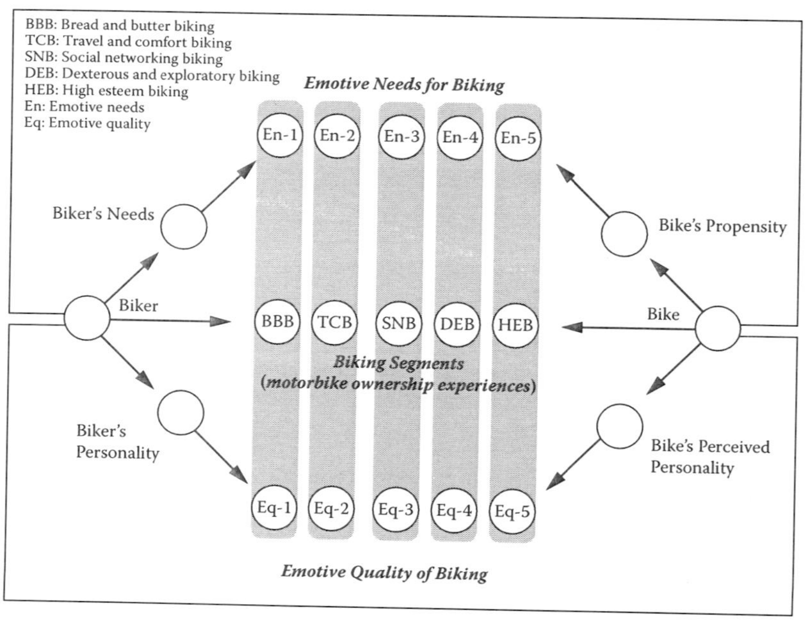

FIGURE 6.3
Emotion-centered model for motorbike styling in India.

(En) and emotive quality (Eq) associated with each biking segment. The model explains the role of biker and bike in shaping a motorbike ownership experience. The model associates the difference in the emotive needs of biking segments to the diversity in bikers' needs and in the bikes' propensity to satisfy these needs. The difference in the emotive quality of different biking segments is also associated to the difference in biker's and bike's personality in each segment.

6.2.1 Concept of *Emoha*

The developed model for motorbike styling was adapted to form a generic model for product styling that has been termed as model for *emoha*-based product styling. The roots of the word *emoha* are derived from three words: (1) "emotion." (2) *moh* (Hindi word meaning *attachment*), (3) *Ha!* (an exclamation of triumph or discovery). *Emoha* refers to the resonance between the emotive needs of a person and the emotive quality of a product and thereby a resonant ownership experience. It is the *emoha-sense* between a person and a product that produces the emotional compatibility that generates a particular product ownership experience. Styling of products on the basis of the *emoha-sense* between a person and a product has been termed as *emoha-based product styling.*

And the methodology associated with *emoha*-based product styling, in the present study of motorbikes, was developed in form of a design method for *emoha*-based product styling. The developed model and design method together represent an emotion-centered research framework for design innovation that can potentially be applied to diverse product categories that are categorized by cultural segments.

6.2.2 Model for *Emoha*-Based Design Innovation

Figure 6.4 illustrates the model for *emoha*-based product styling. The model divides a product ownership experience into two components: emotive needs for product ownership experience and emotive quality of a product ownership experience. The resonance between these two components creates a mutually reinforcing compatibility between the person and the product. This resonance between the two components has been termed *emoha.*

Figure 6.5 shows two different arrangements of the *emoha* model. Figure 6.5a shows the coming together of the emotive needs and the emotive quality, whereas Figure 6.5b shows them going away from each other. The coming together of emotive needs and emotive quality refers to the strong *emoha-sense* between these two components, thereby leading to a stronger compatibility between the person and the product. On the other hand, the large distance between the emotive needs and emotive quality in Figure 6.5B refers to the weak *emoha-sense* between these two components, thereby leading to lesser compatibility between the person and the product.

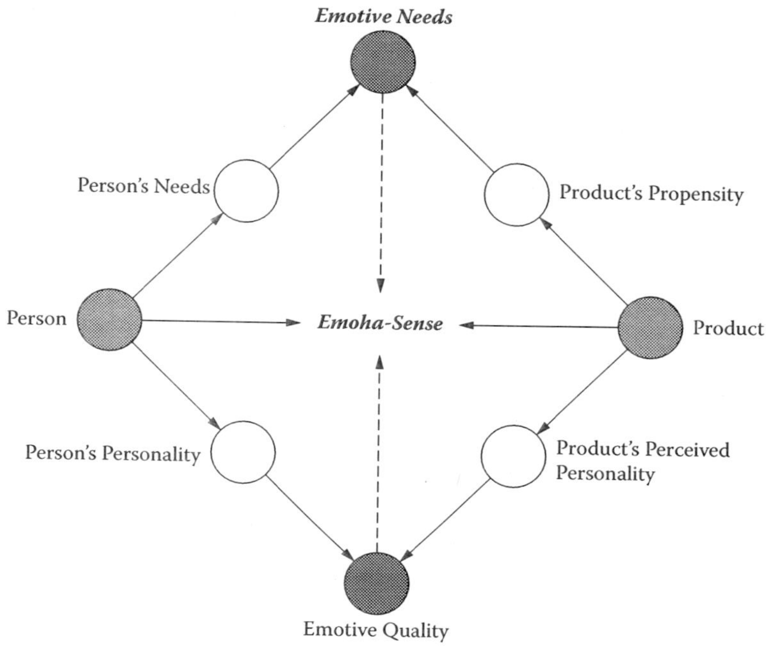

FIGURE 6.4 (See color insert)
Model for *emoha*-based product styling.

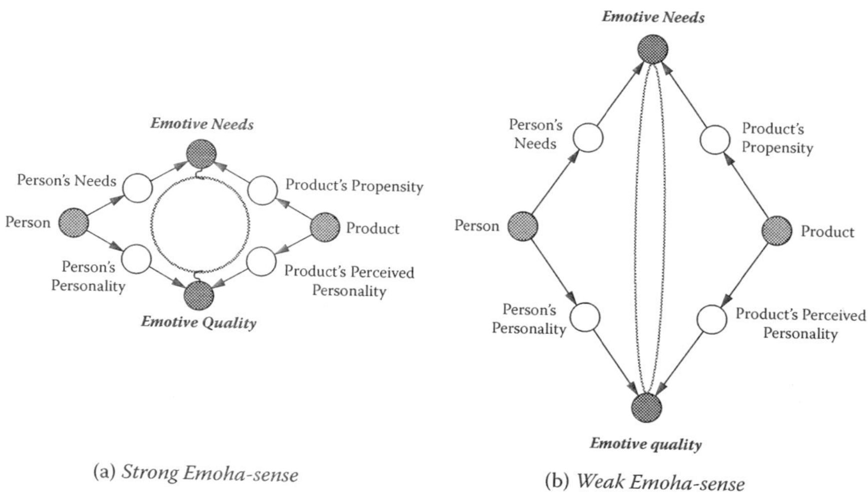

FIGURE 6.5 (See color insert)
Depiction of strong and weak *emoha-sense*.

The *emoha-sense* is represented in form of the flexible band in Figure 6.5 that can be maintained in its original shape only if there is a resonance between the emotive needs and emotive quality. Stretching of the band signifies the loss of resonance, which beyond a point can lead to a rejection of the product ownership experience (band breaks).

6.2.3 Design Method for *Emoha*-Based Product Innovation

The methodology adopted to investigate the emotive quality of motorbike ownership experiences was one of the significant contributions of the present study. This has been presented in form of a design method that can potentially be applied to diverse product categories. The design method for using *emoha*-based product styling in the context of a styling studio is presented in Table 6.1.

Designers often use keywords and associated images in form of image boards, mood boards, and so forth, to visually represent the emotive quality desired in a product. The efficacy of these image-based tools for emotion-centered styling largely relies on the appropriateness of the keywords selected by the designer. It is seen that most of the times the selection of these keywords is based on the personal judgment of designers and market researchers. An element of uncertainty is associated with this personal judgment. The innovative method of designing outlined in Table 6.1 reduces this uncertainty by providing a procedure for selecting keywords in the form of personality variables associated with a product ownership experience.

The model and design method explained earlier together represent an emotion-centered research framework for product styling and design innovation. Integrating this research framework into product design and innovation process would certainly enhance designers' competence for styling products for different ownership experiences.

6.3 Significance of the Study

The following points recapitulate the significance of this study for product designers and mass manufacturers:

1. This study proposes an approach to establish an "emotion-centered segmentation" of product ownership experiences. Such segmentation will facilitate effective communication of the users' needs and styling preferences among all the stakeholders of the design and development process.
2. Knowledge of the emotive quality of mass-manufactured products gives design directions (in form of emotive keywords) for styling

TABLE 6.1

Design Method for *Emoha*-Based Product Innovation

Design Method for *Emoha*-Based Product Innovation
Aim
To evaluate the emotive quality of ownership experiences for a product in a given region to facilitate emotion-centered product styling.
Outline
Step 1: For a given product, identify the set of personality variables that characterize it in the chosen domain of study.
Identify the most popular distinct adjectives that are used to describe the products on related websites, community forums, promotional materials, advertisements, etc. These adjectives will be used a personality variables.
Reference in current study: Chapter 3, Section 3.1
Step 2: Develop a person–product personality measurement instrument.
Use the identified personality variables to develop a self-reporting instrument that can be used for investigating person–product personality for each variable on a five-point rating scale.
Reference in current study: Chapter 3, Section 3.1
Step 3: Conduct person–product personality survey of sample population.
Conduct a person–product personality assessment survey of a sample population using on the developed instrument.
Reference in current study: Chapter 3, Sections 3.2, 3.3, and 3.4
Step 4: Identify the clustering of person–product personality variables based on factor analysis.
Investigate the person–product personality for the sample population. Identify the clusters of correlated person–product personality variables (personality factors) using the statistical method of factor analysis.
Reference in current study: Chapter 5, Sections 5.1, 5.2, and 5.3
Step 5: Identify group of sample population for whom the product is to be designed and evaluate the emotive quality of the ownership experience desired by the selected group.
Select a group of sample population for whom the product is to be designed. Evaluate the significance (factor value score) of the identified person–product personality factors for this group.
Reference in current study: Chapter 5, Sections 5.4, 5.5, and 5.6
Step 6: Create image boards as a styling brief for representing the emotive quality of the ownership experience desired by the selected group.
The person–product personality factors are visually represented in form of image boards with due importance given to the relative significance of each factor for the chosen group. These boards provide visual cues to the designers for building the desired emotive quality of the ownership experience.
Step 7: *Emoha*-based design innovation.
Use the image boards as a styling brief for new product concepts that will provide the desired ownership experience.
Application
Figure 6.6 shows the application of this design method for innovating the designing of motorbikes in the context of the present study of motorbikes in India.

new product versions or innovating existing models for a particular ownership experience. The emotive quality of a product concept can also be investigated through research to evaluate the appropriateness of the proposed concept for the target ownership experience.

3. Mass manufacturers can use the developed research framework to track the changes in styling preferences by monitoring the emotive quality of product ownership experience at different intervals of time. This will enable manufacturers to identify the extent of styling changes required in different models, thereby aiding in development of future product lineups and allocation of resources for product development.
4. The developed research framework facilitates methodical assessment of product styling rather than arbitrary interpretation of design by all stakeholders. This also allows for seamless communication of the emotive concept of product models to advertisement teams and dealer staff that interface with the customers.
5. The developed research framework can be used to compare the emotive quality of products belonging to the same market segment. Such a comparison can give an in-depth knowledge of the reasons for a styling failure or success.
6. The developed research framework can also be used for designing products for unexplored market segments. This can be done by investigating the emotive quality of other popular products being used by potential first-time users.
7. And finally, the study opens up avenues for product designers to conduct independent design research that can complement the conventional research conducted by marketing departments of most product manufacturers.

The present study, through its exclusive focus on the role of emotions in designing a product, will facilitate research orientation among practicing designers both within the Indian industry and elsewhere. Also, by doing so, it would make the design discipline more innovative and amenable to market realties as well as requirements of the owners.

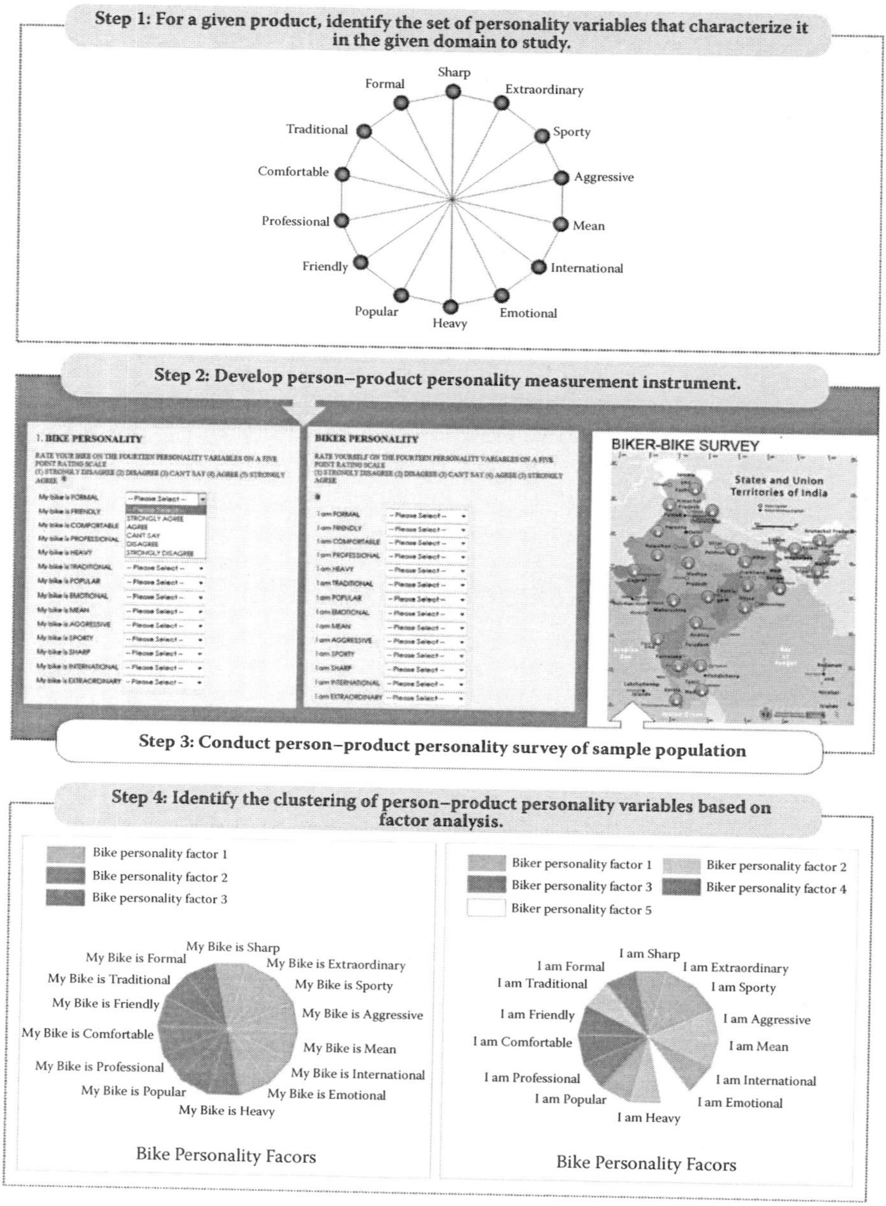

FIGURE 6.6 (See color insert)
Application of the "design method for *emoha*-based product styling" in the context of present study of motorbikes in India.

Continued

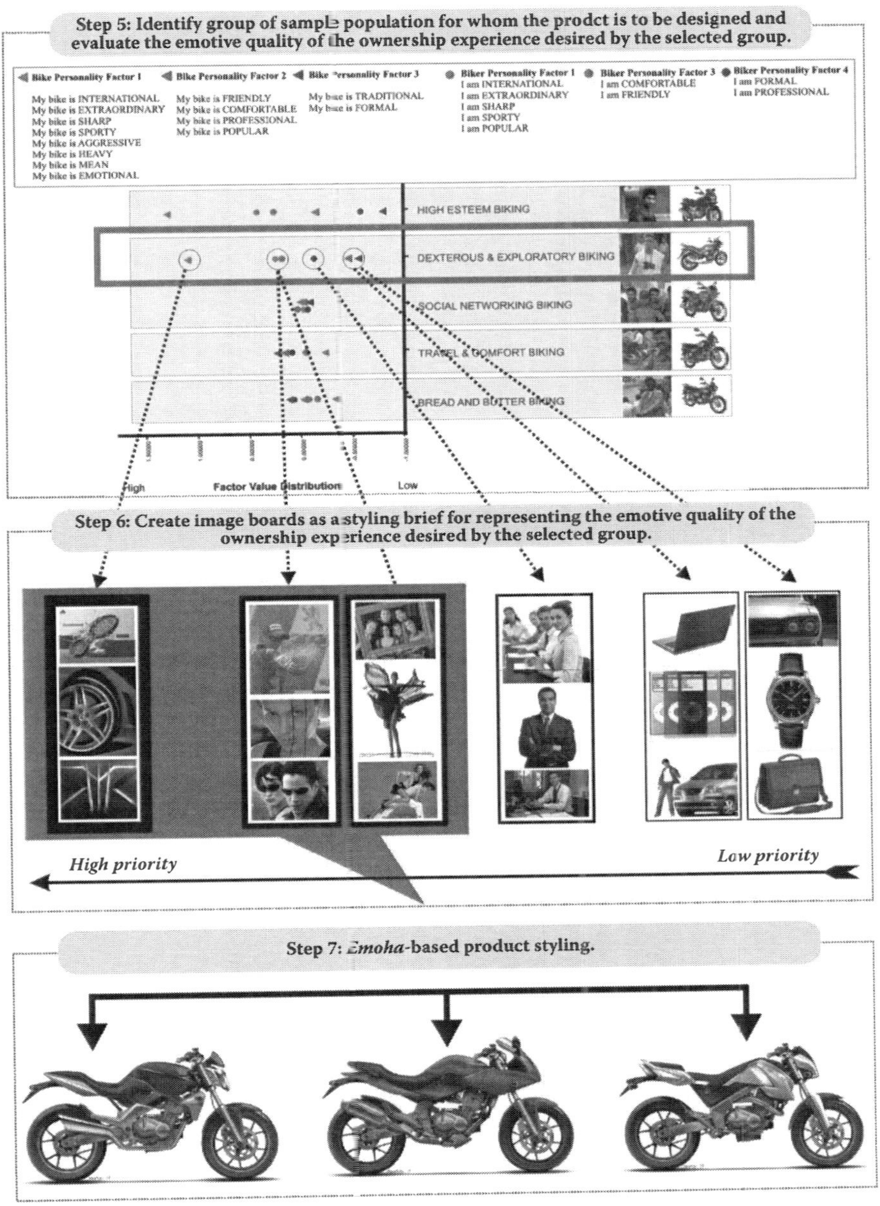

FIGURE 6.6 (*Continued*) (See color insert)
Application of the "design method for *emoha*-based product styling" in the context of present study of motorbikes in India.

References

AMP (2006). *Automobile Mission Plan 2006–2016*. Department of Heavy Industries, Government of India. Retrieved April 22, 2009, from http://www.dhi.nic.in/draft_automotive_mission_plan.pdf.

Angyal, A. (1941). *Foundations for a Science of Personality*. Cambridge, MA: Harvard University Press.

Appadurai, A. (1986). Introduction: Commodities and the politics of value. In Appadurai, A. (Ed.), *The Social Life of Things: Commodities in Cultural Perspective*, 3–63. Cambridge: Cambridge University Press.

Arnold, M. B. (1960). *Emotion and Personality*. New York: Columbia University Press.

Aslam, M. M. (2006). Are you selling the right colour? A cross-cultural review of colour as a marketing cue. *Journal of Marketing Communications*, 12(1), 15–30.

Attfield, J. (2000). *Wild Things: The Material Culture of Everyday Life*. Oxford: Berg Publishers.

Averill, J. R. (1980). Constructivist view of emotions. In R. Plutchik and H. Kellerman (Eds.), *Emotions: Theory, Research and Experience*, Vol. 1, 305–339. New York: Academic Press.

Aydmh, S., Eren, C., Erkok, F., and Sonmez, F. U. (2004). Bodily experience in space and design: The case of a first year studio project at ITU Department of Architecture. *Proceedings of the Fourth International Conference on Design and Emotion* [CD-ROM], July 12–14, 2004, Department of Industrial Design, Middle East Technical University, Ankara, Turkey.

Bahauddin, A., and Abdullah, A. (2004). The Songket motiks—The design and memory of the Malay people. *Proceedings of the Fourth International Conference on Design and Emotion* [CD-ROM], July 12–14, 2004, Department of Industrial Design, Middle East Technical University, Ankara, Turkey.

Barone, M. J., Shimp, T. A., and Sprott, D. E. (1999). Product ownership as a moderator of self-congruity effects. *Marketing Letters, 10* (1), 75–85.

Beaglehole, E. (1932). *Property: A Study in Social Psychology*. New York: Macmillan.

Belk, R. W. (1988). Possessions and the extended self. *Consumer Research, 15*, 139–168.

Ben-Peshat, M. (2004). Popular design and cultural identities: Emotional exchange: Study cases in Israel. *Proceedings of the Fourth International Conference on Design and Emotion* [CD-ROM], July 12–14, 2004, Department of Industrial Design, Middle East Technical University, Ankara, Turkey.

Berger, P. L., and Luckmann, T. (1966). *The Social Construction of Reality: A Treatise It's the Sociology of Knowledge*. Garden City, New York: Anchor Books.

Blattberg, R., Peacock, P., and Sen, S. (1976). Purchasing strategies across product categories. *Journal of Consumer Research, 3*, 143–54.

BMW (2009). *BMW Motorrad: Perfect Riding Pleasure*. Retrieved April 22, 2009, from http://www.bmwmotorcycles.com/us/en/index.html

BMW-HP2 Sport (2009). *BMW HP2 Sport Uncompromisingly Athletic*. Retrieved April 22, 2009, from http://wwwbmw-motorrad.com/com/en/index.html?content = http://www.bmw-motorrad.com/com/en/bikes/highperformance/hp2s/hp2s_main.html

Boradkar, P. (2004). Skin care: The sensual surfaces of objects. *Proceedings of the Fourth International Conference on Design and Emotion* [CD-ROM], July 12–14, 2004, Department of Industrial Design, Middle East Technical University, Ankara, Turkey.

Borja de Mozota, B. (2004). *Design Management: Using Design to Build Brand Value.* New York: Allworth Press.

Botchen, G., Thelen, E., and Pieters, R. (1999). Using means-end structures for benefit segmentation. *European Journal of Marketing, 33* (1/2), 38–58.

Chan, P. K., and Sanders, B.-N.E. (2004). Exploring, interpreting, and applying emotional-driven design in brand identity development: A design student case study. *Proceedings of the Fourth International Conference on Design and Emotion* [CD-ROM], July 12–14, 2004, Department of Industrial Design, Middle East Technical University, Ankara, Turkey.

Chang, W. C. and Wu, T. Y. (2004). Exploring types and characteristics of product forms in eliciting a user's pleasure. *Proceedings of the Fourth International Conference on Design and Emotion* [CD-ROM], July 12–14, 2004, Department of Industrial Design, Middle East Technical University, Ankara, Turkey.

Chhibber, S., Porter, C. S., Porter, J. M., and Healey, L. (2004). Designing pleasure: Designers' needs. *Proceedings of the Fourth International Conference on Design and Emotion* [CD-ROM], July 12–14, 2004, Department of Industrial Design, Middle East Technical University, Ankara, Turkey.

Collopy, F. (2000). Colour, form, and motion: Dimensions of a musical art of light. *Leonardo, 33*(5), 355–360.

Cotte, J., and Robin, R. (2005). Advertisers' theories of consumers: Why use negative emotions to sell? In Geeta Menon and Akshay R. Rao (Eds.), *Advances in Consumer Research*, 24–31. Duluth, MN: Association for Consumer Research.

Crilly, N., Moultrie, J., and Clarskson, P. John (2004). Understanding products appearance: Cognitive and emotional response to products visual form. *Proceedings of the Fourth International Conference on Design and Emotion* [CD-ROM], July 12–14, 2004, Department of Industrial Design, Middle East Technical University, Ankara, Turkey.

Dant, T. (1999). *Material Culture in the Social World: Values, Activities, and Lifestyles.* Buckingham, UK: Open University Press.

Das, L. K. (2002). Towards a non-parochial, non-partisan framework for the study of design history. *Presented at the 3rd International Conference on Design History and Design Studies*, July 9–12, 2002, Istanbul, Turkey.

de Chernatony, L., and de Mello S. C. B. (1995). Predicting brand preferences using self-concept theory. *Journal of Marketing Communications, 1*, 121–135.

De Rivera, J. (1977). *A Structural Theory of the Emotions.* New York: International Universities Press.

Desmet, P. M. A. (2002). *Designing Emotions.* Delft, the Netherlands: Delft University of Technology.

Desmet, P. M. A. (2004). A typology of product emotions. *Proceedings of the Fourth International Conference on Design and Emotion* [CD-ROM], July 12–14, 2004, Department of Industrial Design, Middle East Technical University, Ankara, Turkey.

Desmet, P. M. A. (2008a). Inspire and desire. In P. M. A. Desmet, M. A. Karlsson, and J. van Erp. (Eds.), *Design and Emotion Moves*, 108–125. Cambridge: Cambridge Scholars Publishing.

Desmet, P. M. A. (2008b). Product emotion. In H. N. J. Schifferstein and P. Hekkert (Eds.), *Product Experience*, 379–397. Amsterdam: Elsevier.

Desmet, P. M. A., Ortiz Nicolas, J. C., and Schoormans, J. P. (2008). Product personality in physical interaction. *Design Studies, 29*, 458–477.

Dewey, J. (1934). *Art as Experience*. New York: Perigee Books, the Berkeley Publishing Group.

Dichter, E. (1960). *The Strategy of Desire*. London: Boardman & Co.

Dittmar, H. (1992). *The Social Psychology of Material Possessions: To Have Is to Be*. New York: St. Martin's Press.

Dolnicar, S. (2004). Beyond commonsense segmentation: A systematics of segmentation approaches in tourism. *Journal of Travel Research, 42*, 244–250.

Douglas, M., and Isherwood, B. (1978). *The World of Goods: Towards Anthropology of Consumption*. New York: W. W. Norton.

Ducati (2009). *An Introduction to the Company*. Retrieved April 2009, from http://www.ducati.com/company/introduzione/introduzione.jhtml.

Ducati-1098 (2009). *Ducati Superbike 1098-R*. Retrieved April 22, 2009, from http://www.ducati.com/en/bikes/my2009/ModelPage.jhtml%3Bjsessionid=ZZXKYNROY252GCRNCB2SFEYKFUIHUIV3?family=Superbike&model=SBK1098R-09

Dyne, L. V., and Pierce, J. L. (2004). Psychological ownership and feelings of possession: Three field studies predicting employee attitudes and organizational citizenship behavior. *Journal of Organizational Behavior, 25*, 439–459.

Edwin, Chee Siong L. O. W. (2006). Designing product character: Strategy to evaluate product preference and map design direction. *Presented at the International Conference in Lisbon, IADE. Design Research Society*, November 1–4, 2006, Lisbon. Retrieved April 10, 2009, from http://www.iade.pt/drs2006/wonderground/proceedings/fullpapers/DRS2006_0079.pdf.

Egmond, R. van, Desmet, P. M. A., and Van Der Helm, A. (2004). Basic and cognitive emotions in sound perception. *Proceedings of the Fourth International Conference on Design and Emotion* [CD-ROM], July 12–14, 2004, Department of Industrial Design, Middle East Technical University, Ankara, Turkey.

Ekman, P., and Friesen, W. V. (1978). *Facial Action Coding System: A Technique for the Measurement of Facial Movement*. Palo Alto, CA: Consulting Psychologists Press.

Ekman, P., Friesen, W. V., and Ellsworth, P. (1982). Research foundations. In P. Ekman (Ed.), *Emotion in the Human Face*, 2nd ed., 1–143. New York: Cambridge University Press.

Enquist, H. (2005). Bridging the gap between clinical and patient-provided images. *Presented at the First Conference on Visual Literacy*, April 14–15, 2005, Cork, Ireland. Retrieved April 10, 2009, from http://*www.certec.lth.se/doc/bridgingthegap/bridgingthegap.pdf*

Feagin, J., Orum, A., and Sjoberg, G. (Eds.). (1991). *A Case for Case Study*. Chapel Hill: University of North Carolina Press.

Flyvbjerg, B. (2006). Five misunderstandings about case study research. *Qualitative Inquiry*, 12(2), 219–245.

Forlizzi, J., Multlu, B., and DiSalvo, C. (2004). A study of how products contribute to the emotional aspects of human experience. *Proceedings of the Fourth International Conference on Design and Emotion* [CD-ROM], July 12–14, 2004, Department of Industrial Design, Middle East Technical University, Ankara, Turkey.

Frijda, N. (1987). *The Emotions*. New York: Cambridge University Press.

Fuller, D., Hanlan, J., and Wilde, S. J. (2005). Market segmentation approaches: Do they benefit destination marketers? *Center for Enterprise Development and Research Occasional Paper*, no. 4, Southern Cross University, Coffs Harbour, NSW. *Retrieved July 2, 2009, from http://www.cedar.net.au/assets/pdf/Paper_4.pdf.*

Furby, L. (1978). Possession in humans: An exploratory study of its meaning and motivation. *Social Behavior and Personality, 6*, 49–65.

Galtung, J. (1980). The basic needs approach. In K. Lederer (Ed.), *Human Needs: A Contribution to the Current Debate*, 55–125. Cambridge, MA: Oelgeschlager, Gunn and Hain.

Goldman, A. H. (1995). *Aesthetic Value*. Oxford: Westview Press.

Gomez, R., Popovic, V., and Bucolo, S. (2004). Driving: The emotional experience and automotive design. *Proceedings of the Fourth International Conference on Design and Emotion* [CD-ROM], July 12–14, 2004, Department of Industrial Design, Middle East Technical University, Ankara, Turkey.

Gonzalez, A. M., and Bello, L. (2002). The construct "lifestyle" in market segmentation: The behaviour of tourist consumers. *European Journal of Marketing, 36* (1/2), 51–85.

Gordon, R. M. (1974). The aboutness of emotions. *American Philosophy Quarterly, 11*, 17–36.

Govers, P. C. M., Hekkert, P., and Schoormans, J. P. L. (2004). Happy, cute and touch: Can designers create a product personality that consumers understand? In D. Mc Donagh, P. Hekkert, J. van Erp, and D. Gyi (Eds.), *Design and Emotion: The Design of Everyday Things*, 345–349. London: Taylor & Francis.

Govers, P. C. M., and Mugge, R. (2004). "I love my jeep, because it's tough like me," The effect of product-personality congruence on product attachment. *Proceedings of the Fourth International Conference on Design and Emotion* [CD-ROM], July 12–14, 2004, Department of Industrial Design, Middle East Technical University, Ankara, Turkey.

Gray, J. A. (1982). *The Neuropsychology of Anxiety*. Oxford: Oxford University Press.

Greengrove, K. (2002). Needs-based segmentation: Principles and practice. *International Journal of Market Research, 44* (4), 405–421.

Grubb, E., and Hupp, G. (1968). Perceptions of self, generalised stereotypes, and brand selection. *Journal of Marketing Research, 5*, 58–63.

Grubb, E. L., and Grathwohl, H. L. (1967). Consumer self-concept, symbolism and market behavior: A theoretical approach. *Journal of Marketing, 31*, 22–27.

Gummesson, E. (1991), *Qualitative Methods in Management Research*, 2nd rev. ed. Thousand Oaks, CA: Sage.

Haley, R. (1968). Benefit segmentation: A decision-oriented research tool. *Journal of Marketing, 32* (3), 30–35.

Hanington, B. M. (2004). Death and catharsis: Re-defining pleasure by design. *Proceedings of the Fourth International Conference on Design and Emotion* [CD-ROM], July 12–14, 2004, Department of Industrial Design, Middle East Technical University, Ankara, Turkey.

Harley-Davidson (2009). *Harley-Davidson Inc. 2008 Annual Report*. Retrieved April 22, 2009, from http://www.harley-davidson.com/wcm/Content/Pages/Company/company.jsp? locale = en_US

Harley-Davidson Nightster (2009). *Harley-Davidson Nightster*. Retrieved April 22, 2009, from http://www.harley-davidson.com/wcm/Content/Pages/2009_Motorcycles/2009_Motorcycles.jsp?locale = en_US#/model/xl1200n

Harris, L., and Young, S. (1983). Buyer motivations = human needs: Selling is simpler when you understand the reasons for buying. *Real Estate Today*, (June), 29–30.

Hassenzahl, M., Platz, A., Burmester, M., and Lehner, K. (2000). Hedonic and ergonomic quality aspects determine a software's appeal. In T. Turner and G. Szwillus (Eds.), *Proceedings of the CHI 2000 Conference on Human Factors in Computing*, 201–208. New York: ACM, Addison-Wesley.

Henderson, J. M., Williams, C. C., Castelhano, M. S., and Falk, R. J. (2003): Eye movements and picture processing during recognition. *Perception and Psychophysics, 65* (5), 725–734.

Hero Honda (2009). *Sales Performance: Hero Honda Motors Limited.* Retrieved April 22, 2009, from http://www.herohonda.com/media_sales_performance.asp.

Hilton, K. H. (2004). Attraction considerations. *Proceedings of the Fourth International Conference on Design and Emotion* [CD-ROM], July 12–14, 2004, Department of Industrial Design, Middle East Technical University, Ankara, Turkey.

Hoek, J., Gendall, P., and Esslemont, D. (1996). Market segmentation: A search for the Holy Grail? *Journal of Marketing Practice, 2* (1), 1355–2538.

Hollbrook, M. B. and O'Shaughnessy, J. (1984). The role of emotion and advertising. *Psychology and Marketing, 1*(2), 45–64.

Honda (2009). *Honda Philosophy: Honda's Basic Principles and Policies.* Retrieved April 22, 2009, from http://world.honda.com/profile/philosophy/.

Honda-CBR1000RR (2009). *Meet the Top of the Superbike Food Chain.* Retrieved April 22, 2009, from http://powersports.honda.com/2009/cbr1000rr.aspx.

Hummels, C. C. M. (2004). Educating interaction, experience and diversity. In P. Loyd, N. Roozenburg, C. McMahon, and L. Brodhurst (Eds.), *The Changing Face of Design Education: Proceedings of the 2nd International Engineering and Product Design Education Conference, September 2–3, 2004*, 247–254. Delft: Delft University of Technology Faculty of Industrial Design.

Ingram, J., and Annable, L. (2004). "I see you baby, shakin 'that ass": User perceptions of unintentional anthropomorphism and zoomorphism in consumer products. *Proceedings of the Fourth International Conference on Design and Emotion*, [CD-ROM], July 12–14, 2004, Department of Industrial Design, Middle East Technical University, Ankara, Turkey.

Izard, C. E. (1972). *The Face of Emotion*. New York: Appleton-Century-Crofts.

James, W. (1884). What is an emotion? *Mind, 19*, 188–205.

James, W. (1890). *The principles of psychology*. New York: Dover.

Janlert, L. E., and Stolterman, E. (1997). The character of things. *Design Studies, 18*(3), 297–314.

Kaiser, S., and Wehrle, T. (2001). Facial expressions as indicators of appraisal processes. In K. Scherer, A. Schorr, and T. Johnstone (Eds.), *Appraisal Processes in Emotion*, 285–300. Oxford: Oxford University Press.

Kara, A., and Kaynak, E. (1997). Markets of a single customer: Exploiting conceptual developments in market segmentation. *European Journal of Marketing, 31* (11/12), 873–95.

Karana, E. (2004). The meaning of the material: A survey on the role of material in user's evaluation of a design object. *Proceedings of the Fourth International Conference on Design and Emotion* [CD-ROM], July 12–14, 2004, Department of Industrial Design, Middle East Technical University, Ankara, Turkey.

Karjalainen, T. M. (2004). *Semantic Transformations in Design: Communicating Strategic Brand Identity through Product Design References*. Helsinki: University of Art and Design Helsinki.

Karjalainen, T. M. (2007). It looks like a Toyota: Educational approaches to designing for visual brand. *International Journal of Design, 1*(1), 67–81.

Karjalainen, T. M., and Warell, A. (2005). Do you recognise this tea flask? Transformation of brand-specific product identity through visual design cues. *Proceedings of the International Design Congress, IASDR 2005*, Oct 31–Nov 4, 2005, Douliou, Taiwan.

Karlsson, M. A., and Wikstrom, L. (2009). Safety semantics. A tudy on the effect of product expression on user safety behaviour. *Contemporary Ergonomics. Selected papers and an overview of the Ergonomics Society Annual Conference* 1984–2008, pp. 609–613.

Kim, J., and Moon, J. Y. (1998). Designing towards emotional usability in customer interfaces-trustworthiness of cyber-banking system interfaces. *Interacting with Computers, 10*, 1–29.

Kleine, R. E., III, Kleine, S. S., and Kernan, J. B. (1993). Mundane consumption and the self: A social-identity perspective. *Journal of Consumer Psychology, 2(3)*, 209–235.

Krippendorff, K. (2004). *Content Analysis: An Introduction to Its Methodology*. Thousand Oaks, CA: Sage Publications.

Krippendorff, K. (2006). *The Semantic Turn: A New Foundation for Design*. Boca Raton, FL: Taylor & Francis.

Krippendorff, K., and Butter, R. (1984). Product semantics: Exploring the symbolic qualities of form. *Innovation, 3*(2), 4–9.

Kumar, K. L. (2002). Product modeling for aesthetics. *Applied Simulation and Modeling, Proceedings of the IASTED International Conference*, June 25–28, 2002, 31–35. Crete, Greece.

Lang, P. J. (1980). Behavioral treatment and bio-behavioral assessment: Computer applications. In J. B. Sidowski, J. H. Johnson, and T. A. Williams (Eds.), *Technology in Mental Health Care Delivery System*, 129–139. Norwood, NJ: Albex.

Larsen, R. J., and Diener, E. (1992). Promises and problems with the circumplex model of emotion. In M. S. Clark (Ed.), *Emotion*, 25–59. Newbury Park, CA: Sage Publications.

Lazarus, R. S., Averill, J. R., and Opton, E. M., Jr. (1970). Toward a cognitive theory of emotions. In M. Arnold (Ed.), *Feelings and Emotions*, 207–232. New York: Academic Press.

Mallmann, C. (1980). Society, needs and rights. In K. Lederer (Ed.), *Human Needs—A Contribution to the Current Debate*, 37–54. Cambridge, MA: Oelgeschlager, Gunn and Hain.

Mandler, G. (1975). *Mind and Emotion*. New York: Wiley.

Mano, H. (1996). Assessing emotional reactions to TV ads: A replication and extension with a brief adjective checklist. *Advances in Consumer Research, 23*, 63–69.

Maslow, A. (1954). *Motivation and Personality*. New York: Harper & Row.

Maslow, A. (1968). *Towards a Psychology of Being*. New York: Van Nostrand Reinhold.

Max-Neef, M. (1991). *Human-Scale Development: Conception, Application and Further Reflection*. London: Apex Press.

Max-Neef, M. (1994). Development and human needs. In P. Ekins and M. Max-Neef (Eds.), *Real-Life Economics: Understanding Wealth Creation*, 197–213. New York: Routledge.

McDougall, W. (1926). *An Introduction to Social Psychology*. Boston: Luce and Co.

McDonald, M. (1999). *Marketing Plans. How to Prepare Them: How to Use Them*, 4th ed. Oxford: Butterworth Heinemann.

McDonald, M., and Dunbar, I. (1995) *Market Segmentation: A Step-by-Step Approach to Creating Profitable Market Segments*. London: Whitaker.

McDougall, W. (1928). Emotion and feeling distinguished. In Reymert, M. L. (Ed.), *Feelings and Emotions*. Worcester, MA: Clark University Press.

Mehrabian, A., and Russel, J. (1977). Evidence for a three-factor theory of emotions. *Journal of Research in Personality, 11*, 273–294.

Morgan, G., and Smircich, L. (1980). The case for qualitative research. *The Academy of Management Review, 5* (4), 491–500.

Mowrer, O. H. (1960). *Learning and Behavior*. New York: Wiley.

Mugge, S., Hendrik, N. J., and Schoormans, J. P. L. (2004). Personalizing product appearance: The effect on product attachment. *Proceedings of the Fourth International Conference on Design and Emotion* [CD-ROM], July 12–14, 2004, Department of Industrial Design, Middle East Technical University, Ankara, Turkey.

Mugge, R., Schifferstein, H. N. J., and Schoormans, J. P. L. (2006). Product attachment and product lifetime: The role of personality congruity and fashion. In K. M. Ekström, and H. Brembeck (Eds.), *European Advances in Consumer Research, 7*, 1–7.

Myers, J. (1976). Benefit structure analysis: A new tool for product planning. *Journal of Marketing, 40*, 23–32.

Noor, K. B. M. (2008). Case study: A strategic research methodology. *American Journal of Applied Sciences, 5* (11), 1602–1604.

Norman, D. A. (1990). *The Design of Everyday Things*. New York: Doubleday.

Norman, D. A. (2004). *Emotional Design: Why We Love (or Hate) Everyday Things*. New York: Basic Books.

OALD (2005). *Oxford Advanced Learner's Dictionary of Current English*, 7th ed. New York: Oxford University Press. 2005.

Oatley, K., and Johnson-Laird, P. N (1987). Towards a cognitive theory of emotions. *Cognition and Emotion*, 1, 29–50.

Ortony, A., Clore, G. L., and Collins, A. (1988). *The Cognitive Structure of Emotions*. New York: Cambridge University Press.

Osman, M. Z. (1993). A conceptual model of retail image influences on loyalty patronage behavior. *The International Review of Retail, Distribution and Consumer Research, 31*, 149–166.

Oustrup, M. K., and Jeanes, M. (2005). MOOD consumption theory: A human-focused marketing tool. Presented at ESOMAR, Annual Congress, September 18–21, 2005, Cannes.

Panksepp, J. (1982). Toward a general psychobiological theory of emotions. *The Behavioral and Brain Sciences, 5*, 407–467.

Parkinson, B., and Manstead, A. S. R. (1992). Appraisal as a cause of emotion. In Clark, Margaret S. (Ed.), *Emotion* (pp. 122–149). Newbury Park, CA: Sage Publications.

Paulhan, F. (1884). *The Laws of Feeling* (C. K. Ogden, Trans.). New York: Harcourt, Brace. (Original work published in 1971.)

Pham, B. (1999). Design for aesthetic: Interactions of design variables and aesthetic properties. *Proceeding of SPIE IS and T/SPIE 11th Annual Symposium-Electronic Imaging'99*, 364–371. Retrieved April 5, 2009, from http://eprints.qut.edu.au/2047/1/2047_2.pdf

Plutchik, R. (1980). A general psychoevolutionary theory of emotion. In R. Plutchik and H. Kellerman (Eds.), *Emotions: Theory, Research, and Experience*, Vol. 1, 3–33. New York: Academic.

Porter, C. S., Chhibber, S., and Porter, J. M. (2003). The development of an inclusive design resource. *Proceedings of INCLUDE 2003: Inclusive Design for Society and Business* [CD-ROM], March 25–28, 2003 (578–585). London: Royal College of Art.

Porter, S., Chhibber, S., and Porter, J. M. (2002). Towards an understanding of pleasure in product design. In D. Mc Donagh, P. Hekkert, J. van Erp, and D. Gyi (Eds.), *Design and Emotion: The Design of Everyday Things*, 291–295. London: Taylor & Francis.

Pribram, K. H., and Melges, F. T. (1969). Psychophysiological basis of emotion. In P. J Vinken, and G. W. Brunyn (Eds.), *Handbook of Clinical Neurology*, 316–342. Amsterdam: North Holland.

Prown, D. J. (1982). Mind in matter: An introduction to material culture theory and method. *Winterthur Portfolio, 17*(1), 1–19.

Ribas, V. G., Battaiola, A. L., Montenegro, B. C., Andrade, J. F., and Queiroz, Yuri C. V. (2004). Reference board: Method with focus on emotion for the conception of new products in design. *Proceedings of the Fourth International Conference on Design and Emotion* [CD-ROM], July 12–14, 2004, Department of Industrial Design, Middle East Technical University, Ankara, Turkey.

Richins, M. L. (1994a). Valuing things: The public and private meanings of possessions. *Journal of Consumer Research, 21* (December), 504–521.

Richins, M. L. (1994b). Special possessions and the expression of material values. *Journal of Consumer Research, 21* (December), 522–533.

Russell, J. A. (1980). A circumplex model of affect. *Journal of Personality and Social Psychology, 39*, 1161–1178.

Russell, K., et al (2004) Artefacts and affordance: The surface of meaning. *Working Papers in Art and Design* 3. Retrieved July 3, 2009, from http://www.herts.ac.uk/artdes/research/papers/wpades/vol3/krfull.html.

Sahlins, M. (1976). *Culture and Practical Reason*. Chicago: University of Chicago Press.

Sartre, J.-P. (1948). *The Emotions: Outline of a Theory*. New York: Philosophical Library.

Savas, O. (2004). 'Design and emotion' in the context of social differentiation. *Proceedings of the Fourth International Conference on Design and Emotion* [CD-ROM], July 12–14, 2004, Department of Industrial Design, Middle East Technical University, Ankara, Turkey.

Schachter, S., and Singer, J. (1962). Cognitive, social and physiological determinants of emotional state. *Psychology Review, 69*, 379–399.

Schifferstein, H. N. J. (2004). Sensing the senses: Multimodal research with applications in product design. *Proceedings of the Fourth International Conference on Design and Emotion* [CD-ROM], July 12–14, 2004, Department of Industrial Design, Middle East Technical University, Ankara, Turkey.

Schiffman, L., Bednall, D., Cowley, E. E., O'Cass, A., Watson, J., and Kanuk, L. (2001). *Consumer Behaviour*, 2nd ed. Pearson Education Australia Pty Ltd. Australia.

Schmitt, B., and Simonson, A. (1997). *Marketing Aesthetics: The Strategic Management of Brands, Identity and Image*. New York: Free Press.

Sevener, Z., and Asatekin, M. (2004). Influence of aesthetic properties on stimulating emotional responses. *Proceedings of the Fourth International Conference on Design and Emotion* [CD-ROM], July 12–14, 2004, Department of Industrial Design, Middle East Technical University, Ankara, Turkey.

SIAM (2009). *Society of Indian Automobile Manufacturers.* Retrieved January 2009 from http://www.siamindia.com/.

Sirgy, M. J. (1982). Self-concept in consumer behaviour: A critical review, *Journal of Consumer Research, 9*, 287–300.

Smith, W. (1956). Product differentiation and market segmentation as alternative marketing strategies. *Journal of Marketing, 21*(1), 3–8.

Stompff, G. (2003). The forgotten bond: Brand identity and product design. *Design Management Journal, 14* (1), 26–32.

Tomkins, S. S. (1962). *Affect Imagery Consciousness. The Positive Affects,* Vol. 1. New York: Springer.

Tomkins, S. S. (1984). Affect theory. In K. R. Scherer and P. Ekman (Eds.), *Approaches to Emotion*, 163–195. Hillsdale, NJ: Erlbaum.

Tsai, H. C., Hung, C. Y., and Hung, F. K. (2007). Computer aided product color design with artificial Intelligence. *Computer-Aided Design and Applications, 4* (1–4), 557–564.

Tunnicliff, D. J. (2006). *Psychosocial Factors Contributing to Motorcyclists' Intended Riding Style: An Application of an Extended Version of the Theory of Planned Behaviour* [QUT Book]. Retrieved April 15, 2009, from http://eprints.qut.edu.au/16217/.

Vuori, T., Olkkonen, M., Polonen, M. Siren, A., and Hakkinen, J. (2004): Can eye movements be quantitatively be applied to image quality studies? In Raisamo, R. (Ed.), *NordiCHI 2004—Proceedings of the Third Nordic Conference on Human-Computer Interaction*. Association for Computing Machinery.

Watson, D., and Tellegen, A. (1985). Toward a consensual structure of mood. *Psychological Bulletin, 3*, 1–14.

Watson, J. B. (1930). *Behaviorism*. Chicago: University of Chicago Press.

Webb, B., and Gountas, J, (2006). An integrative model of brand personality, self–concept and consumer personality orientations. *Australian and New Zealand Marketing Academy (ANZMAC) Conference,* December 4–6, 2006, Queensland University of Technology, Brisbane. Retrieved January 16, 2008, from http://mib.vuw.ac.nz:8081/WWW/ANZMAC2006/documents/Webb_Ben.pdf.

Weiner, D., and Graham, S. (1984). An attributional approach to emotional development. In C. E. Izard, J. Kagan, and R. B. Zajonc (Eds.), *Emotions, Cognition, and Behavior*, 167–191. New York: Cambridge University Press.

Wilpert, B. (1991). Property, ownership, and participation: On the growing contradictions between legal and psychological concepts. In R. Russell and V. Rus (Eds.), *International Handbook of Participation in Organizations: For the Study of Organizational Democracy, Co-operation, and Self Management*, Vol. 2, 149–164. New York: Oxford University Press.

Wind, Y. (1978). Issues and advances in segmentation research. *Journal of Marketing Research, 15*, 317–338.

Yamaha (2009). *Yamaha Corporate Philosophy. Creating Kando from Products and Services That Exceed Your Expectations.* Retrieved April 22, 2009, from http://www.yamaha-motor.co.jp/global/about/philosophy/.

Yamaha-YZF-R6 (2009). *Yamaha YZF-R6- Catch the Extreme.* Retrieved April 22, 2009, from http://www.yamaha-motor-europe.com/products/motorcycles/supersport/yzf-r6.jsp.

Yankelovich, D. (1964). New criteria for market segmentation, *Harvard Business Review*, 42, 83–93. Retrieved May 5, 2009, from http://www.*danyankelovich.com/newcriteria.pdf.*

Yin, R. K. (1984). *Case Study Research: Design and Methods.* Beverly Hills, CA: Sage Publications.

Zainal, Z. (2007). Case study as a research method. *Jurnal Kemanusiaan*, *9*, 1–6.

Zimmerman, J., and Ayoob, E. (2004).The role of products in consumer-celebrity relationships. *Proceedings of the Fourth International Conference on Design and Emotion*, [CD-ROM], July 12–14, 2004, Department of Industrial Design, Middle East Technical University, Ankara, Turkey.

Index

O

P

Q

R

T

U

V

W

X

Z